M000195146

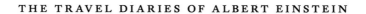

THE TRAVEL DIARIES OF ALBERT EINSTEIN

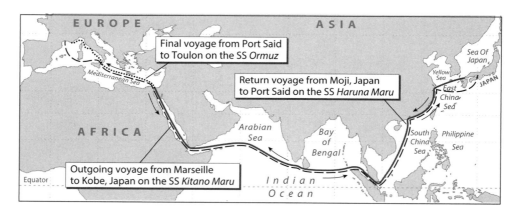

Map with itinerary of Einstein's ocean voyages

THE TRAVEL DIARIES OF
# Albert Einstein

EDITED BY
ZE'EV ROSENKRANZ

THE FAR EAST, PALESTINE & SPAIN
1922–1923

*Albert Einstein,*

PRINCETON UNIVERSITY PRESS
PRINCETON AND OXFORD

Editorial apparatus and diary translation
copyright © 2018 by Princeton University Press

Travel diary, additional texts, and endpaper images
copyright © 2018 by the Hebrew University of Jerusalem

Requests for permission to reproduce material from this work
should be sent to Permissions, Princeton University Press

Published by Princeton University Press
41 William Street, Princeton, New Jersey 08540

In the United Kingdom: Princeton University Press
6 Oxford Street, Woodstock, Oxfordshire OX20 1TR

press.princeton.edu

Cover and book design by Chris Ferrante

All Rights Reserved

ISBN 978-0-691-17441-9

Library of Congress Control Number: 2018931638

British Library Cataloging-in-Publication Data is available

This book has been composed in Kis Antiqua Now

Printed on acid-free paper. ∞

Printed in the United States of America

10 9 8 7 6 5 4 3 2 1

To my parents
Ruth Hannah Rosenkranz (1928–2015)
&
Arnold Rosenkranz (1923–1999)
with whom I embarked on
my very first voyages

# Contents

# Illustrations

# Preface

I GREW UP A FIVE-MINUTE WALK from the beach in the bayside suburb of St. Kilda in Melbourne, Australia. It was the Sixties and early Seventies, and St. Kilda was still a bit seedy—it has since been partially gentrified. The Jewish refugees from Europe would stand on Acland St. and converse in languages I could not understand. My Anglo-Dutch Jewish maternal grandfather—who had himself immigrated to Australia from London in the aftermath of the Great War—would call these more recent immigrants "the refs," a derogatory term for refugees.

My Viennese father loved to take me on a three-mile, hour-long weekly walk along Beaconsfield Parade to Station Pier in Port Melbourne to see the ocean liners that had docked there. He was always drawn to the sea. Born in landlocked Vienna, he had fled to Shanghai in late 1938 and lived there for ten years, including almost four years under Japanese occupation in the Jewish ghetto under harrowing conditions. After the end of World War II, he easily made friends with visiting American sailors. In 1948, as a committed Zionist, he left on the second ship to depart from Shanghai with Jewish refugees for the newly formed State of Israel. But the country could not hold my father (who was the personification of the "Wandering Jew") for long. In 1950, he returned to Vienna, mainly due to homesickness. Six years later, he was traveling by sea again—this time to Australia to start a new life there, just in time to attend the Melbourne Olympics. A few years after that, he met my Melbourne-born mother. Yet his *wanderlust* continued and, when I was almost eleven, we returned to Station Pier, this time as passengers.

The four-week voyage on an Italian ship took the remigrating tailor and his family back to Europe via the Cape of Good Hope (the

Suez Canal was still under blockade following the Six Day War). I found life on the ocean liner somewhat boring; I was too old for the "children's' room" and too young for the adults' activities. But I made some good friends onboard, and that helped pass the time. My most vivid memories from that voyage were the frequent announcements in the ship's *Sea Herald* to set the clocks back by half an hour, as we traversed the globe and changed time zones, and the many times I sneaked into the small cinema on board to watch the Japanese sci-fi movie, *The Green Slime*. I loved the film so much that I would attend both the English and the Italian viewings.

Two more ocean crossings would ensue within the next three years, as my father found it difficult to decide where he wanted to live. It wasn't the most stable lifestyle for his prepubescent son, but coming of age in both Australia and Austria was a good way to pick up languages and an internationalist outlook. Growing up Jewish in the Vienna of the 1970s certainly had its challenges. One of them was coping with occasional anti-Semitic remarks by our *Klassenvorstand*, our class director, at the local *Gymnasium*. One of his favorite bon mots was to refer to the "Affentürkei" ("Turkey of the monkeys"). At the time, I had no idea that this was an anti-Semitic comment, apart from being racist in other ways as well. The xenophobic "joke" in Vienna at the time was that the Orient began just east of the Danube Canal, where, before the Holocaust, most of the Jews, including my father, had lived. But it had also been home to the members of other ethnic groups from the East from the erstwhile Austro-Hungarian empire, mainly Czechs and Slovaks. In 1970s Vienna, it was still home to a small Jewish minority but also more recent arrivals from Yugoslavia and Turkey.

Even though I really took pleasure in those long walks to Port Melbourne, I subconsciously chose to live in "landlocked" cities in my adult life. From Vienna, I moved to Jerusalem, and then eventually to Pasadena in California, both an hour from the nearest beach.

Albert Einstein's travel diaries are, by far, my favorite documents penned by him. I have always enjoyed his quirky style, his acerbic quips about the individuals he met, and the colorful descriptions of the hustle and bustle in his ports of call. It was only later that I started to notice the more troublesome entries in his journal, in which he, at times, made what amounted to xenophobic comments about some of the peoples he encountered. I began to ask myself: how could this humanist icon be the author of such passages?

The answer to this question seems very relevant in today's world, in which the hatred of the Other is so rampant in so many places around the world. Once again, there are destitute refugees desperately trying to flee persecution and violent conflicts, huddled on unseaworthy vessels, sometimes drowning, and—if they do survive—are penned up in barely humane conditions in refugee camps. Once again, parents are frantically trying to send their children to safety, only to endure heart-wrenching and perilous journeys. Once again, foreigners are being scapegoated for local social ills, and walls of cement, barbed wire, and prejudice are being erected to keep them out.

In such a reality, it is especially pertinent to explore how even the poster-child for humanitarian efforts—the United Nations High Commissioner for Refugees once produced a placard brandishing Einstein's image with the ingenious slogan: "A bundle of belongings isn't the only thing a refugee brings to his new country. Einstein was a refugee"—could have held prejudiced and stereotypical beliefs about the members of other nations. It seems that even Einstein sometimes had a very hard time recognizing himself in the face of the Other.

Pasadena, California, October 2017

# Acknowledgments

I EXPRESS MY DEEP APPRECIATION TO Al Bertrand, associate publishing director, at Princeton University Press, for his astute guidance and insightful advice throughout the publication process. I am especially indebted to Diana Buchwald, director of the Einstein Papers Project, for her discerning comments and unstinting and gracious support of this project. I am also grateful to Peter Dougherty, former director of Princeton University Press, and to Jürgen Renn, co-director of the Max Planck Institute for the History of Science in Berlin, for their kind encouragement and sage counsel. Without the many years of the amiable collaboration of Christine Nelson, curator of literary and historical manuscripts at the Morgan Library & Museum in New York, this publication would not have been possible.

It is a pleasure to express my gratitude to my dear colleagues at the Einstein Papers Project: to Barbara Wolff and Nurit Lifshitz for their kind assistance with my archival and library research and for their very helpful advice; to Dennis Lehmkuhl for his cheerful support and for making sure that I avoid committing any scientific howlers; to Emily de Araújo and Jenny James for their very gracious collaboration; and to Jeremy Schneider for his assistance. It has been a delight working with the exceptionally competent production team at Princeton University Press: Terri O'Prey, associate managing editor, Chris Ferrante, designer, and Dimitri Karetnikov, illustration manager; and with Cyd Westmoreland, my copyeditor, whose meticulous professionalism has been outstanding. I am also deeply indebted to Roni Gross and Chaya Becker of the Albert Einstein Archives for their kind help in research and copyright issues; to Thomas F. Glick for his valuable comments; to Mayuko Akagi of Hitotsubashi

University; Anat Banin of the Central Zionist Archives; Joan Bieder of UC Berkeley; Brianna Cregle of Princeton University Library; Shimon Dias of the Japan Academy; Lisa Ginsburg of Los Angeles; the staff at Kodansha Ltd. of Tokyo; Yuki Ogawa of NYK Maritime Museum; H. Ooe of Waseda University; Marilyn Palmeri of the Morgan Library & Museum; the staff at the David Rumsey Map Collection; Ellen Sandberg of Granger Historical Picture Archive; Michael Simonson of the Leo Baeck Institute; Hideki Sugimoto of Osaka University; Jean-Michel de Tarragon of École Biblique; and Juan José Villar Lijarcio of Archivo General de la Administración for their assistance in securing copyright permissions for the facsimiles and illustrations; to Hallie Schaeffer and Kristin Zodrow at Princeton University Press for their helpful collaboration on the production of this publication; and to John Wade, Dan Agulka, Bianca Rios, and Ben Perez at the interlibrary loan department at the Caltech Library for their untiring efforts in catering to my very idiosyncratic document delivery needs.

Thanks are also due to those individuals who were instrumental in the work on the original publication of Einstein's travel diary to Japan, Palestine, and Spain in Volume 13 of *The Collected Papers of Albert Einstein*: Tilman Sauer, for his excellent work on the scientific contents of the diary; Seiyo Abiko, for his splendid research into Japanese historical materials; Yiwen Ma, for her charming assistance with Chinese-language sources; and Masako Ohnuki, for her gracious assistance with Japanese-language sources. Work on that publication of the diary was also assisted by Luis J. Boya, Danian Hu, Herbert Karbach, Adán Sus, and Victor Wei.

Last, but by no means least, I express my immense gratitude to my partner Kimberly Ady for her keen comments and loving support and encouragement.

# THE TRAVEL DIARIES OF ALBERT EINSTEIN

# Historical Introduction

The voyage is splendid. I am enchanted with Japan and the Japanese and am sure that you would be, too. Moreover, such a sea voyage is a splendid existence for a ponderer anyway—like a monastery. Added to it is the caressing heat near the equator. Warm water drips lazily down from the sky and spreads peace and a plantlike dozing— this little letter testifies to it.

*Albert Einstein to Niels Bohr, near Singapore, 10 January 1923*

## This Edition

This new edition presents Albert Einstein's complete travel diary from his half-year long voyage to the Far East, Palestine, and Spain from October 1922 to March 1923.[1] Facsimiles of each of the pages from the original journal will be accompanied by a translation into English. Scholarly annotations will provide identification of the individuals, organizations, and locations mentioned in the diary entries, elucidate obscure references, supply additional information on the events recorded in the diary, and offer details of his travel itinerary not mentioned by Einstein in his journal. Even though the diary constitutes a riveting historical source in which Einstein recorded his immediate impressions, by its very nature it only provides one piece of the puzzle for reconstructing the course of the trip. Therefore, to obtain a more comprehensive picture of the voyage, the annotations draw on the contemporary local press coverage from the various countries Einstein visited, additional sources from his personal papers, diplomatic reports from the time of the trip, and contemporary articles and later memoirs and reminiscences of the participants. The edition will also include supplementary documents authored by Einstein that provide further context for the travel journal: letters and postcards dispatched from the voyage, speeches he gave at various ports of call along his journey, and articles he wrote on his impressions of the countries he visited, both during his trip and in its aftermath.

This journal was first published in its entirety in Volume 13 of *The Collected Papers of Albert Einstein* (*CPAE*) in 2012.[2] For the purposes of this edition, the translation published in the English translation of Volume 13 of the *CPAE* has been substantially revised.

Previously, selected excerpts of the travel diary had been published in translation: the section of the diary pertaining to Einstein's visit to Japan appeared in Japanese in 2001; the Palestine entries in the diary were published in an edited version in Hebrew in 1999; and the entries for the Spanish leg of the journey appeared in English in 1988. A brief excerpt from the diary was also published in 1975.[3]

## The Travel Diary

This journal is one of six travel diaries penned by Einstein. No diary is extant for Einstein's first overseas voyage to the United States in the spring of 1921. Indeed, we do not know whether he kept a travel diary on that trip.[4] The other extant diaries were written during his trip to South America, from March to May 1925, and his three trips to the United States when he visited the California Institute of Technology in Pasadena during the consecutive winter terms of 1930–1931, 1931–1932, and 1932–1933. Even though this amounts to five overseas trips, there are in fact six diaries, as Einstein used two notebooks on his last trip.[5]

The travel diary presented here was written in a notebook that consists of 182 lined pages. The diary entries appear on 81 lined pages. These are followed by 82 blank lined pages and 19 lined pages and one unlined page of calculations.[6] These calculations were written on the verso of this document (i.e., at the back end of the travel diary) and upside down in relation to the diary entries. The calculations are not included in this edition.

The diary provides us for the very first time with gripping insights into Einstein's journey experience. He wrote daily, on occasion accompanying his text with various drawings of interesting sightings, such as volcanoes, boats, and fish. He recorded both immediate impressions of his travel experiences and longer reflections on his readings, the people he met, and the places he visited. He also jotted down his musings on science, philosophy, art, and, occasionally, contemporary world events. The style of his diary was often very detailed, yet written in a quirky and (possibly due to lack of time) telegraphic manner. His observations of the individuals he encountered on his trip were frequently very succinct in nature—he could sum up their personalities and idiosyncrasies in just a few, often humorous or irreverent, words. As for the reason Einstein started to keep a travel diary for this specific trip, we can only speculate. However, it was most likely intended both as a record for himself and as reading matter for his two step-daughters, Ilse and Margot, who remained at home in Berlin.[7] We can be certain that he did not intend it for posterity or for publication.

The history of this journal is an intriguing one. Following Einstein's decision not to return to Germany in the wake of the Nazis' rise to power in January 1933, his son-in-law, the German-Jewish literary critic and editor Rudolf Kayser, removed Einstein's personal papers (including this travel diary) from his apartment in Berlin to the French Embassy there and arranged for them to be transferred to France by diplomatic pouch. From there, they were shipped to Princeton, New Jersey, where Einstein had taken up residence.[8]

In his last will and testament of 1950, Einstein appointed his loyal friend Otto Nathan and his long-term secretary Helen Dukas as trustees of his estate and Nathan as its sole executor. After

Einstein's death in April 1955, Dukas also became the first archivist of his personal papers.

In February 1971, an agreement was signed between the Estate of Albert Einstein and Princeton University Press to publish a comprehensive edition of Einstein's writings and correspondence. This endeavor became *The Collected Papers of Albert Einstein*. Physics professor John Stachel was appointed as editor-in-chief of the enterprise in 1976. A year later, it became evident that Otto Nathan was not satisfied with Stachel as editor. After first suggesting that he be dismissed, Nathan then urged that a triumvirate of editors be appointed that would include Stachel. When Princeton University Press decided to continue to support Stachel as the project's sole editor-in-chief and reject Nathan's demands, the Einstein Estate took legal action. As the parties could not reach an agreement, the matter eventually came before an arbitration judge in New York in the spring of 1980. As a result, the judge decided in favor of the Press. The legal actions and arbitration proceedings were very costly for the Einstein Estate.[9] Therefore, the trustees decided to sell some of the original items in Einstein's personal papers to defray the costs. In early 1980, Otto Nathan asked the New York publisher and bibliophile James H. Heineman of the Heineman Foundation whether he would be interested in purchasing material from the Einstein Archives. The Heineman Foundation had previously purchased a thirty-five-page scientific manuscript by Einstein in 1976. In a "Note for the files," Nathan wrote "that funds might be needed for the payment of legal expenses resulting from the Court actions against Princeton University Press and from the forthcoming arbitration transactions."[10] In October 1980, the Foundation purchased three important essays by Einstein from the Estate and donated them to the Pierpont Morgan Library in New York.[11] In July 1981, the Estate requested appraisals of several Einstein manuscripts (including this travel diary) from two autograph sellers. One company claimed the market value of the journal was $95,000, whereas the other one priced it at merely $6,500.[12] The following month, the Heineman Foundation purchased the travel diary for an unnamed price from the Estate and immediately deposited it in the "Dannie and Hettie Heineman Collection" at the Pierpont Morgan Library (now the Morgan Library & Museum), where the journal remains to this day.[13]

## Background of the Trip

Various decisive factors formed the background to Einstein's extensive tour of the Far and Middle East and Spain in late 1922 and early 1923.[14] Einstein had traveled beyond the confines of Europe for the first time in the spring of 1921, when he toured the United States, thereby accompanying Chaim Weizmann, president of the London-based Zionist Organisation, to raise funds for the planned Hebrew University in Jerusalem and to establish ties with the American scientific community in the wake of World War I.[15] Without explicitly declaring that he was on a mission to disseminate his new theories of relativity after achieving overnight fame in November 1919, by the time he had embarked on his trip to the Far East, Einstein had already lectured on his theories in numerous locations around Germany, as well as in Zurich, Oslo, Copenhagen, Leyden, Prague, and Vienna.[16] On his return trip from the United States, he had also made a brief visit to the United Kingdom.[17] During his first trip to Paris since the war in April 1922, Einstein participated in discussions on relativity at the Collège de France and at the French Philosophical Society.[18]

The initiative that eventually led to Einstein's trip to the Far East, Palestine, and Spain originated with the president of the Kaizo-Sha publishing house in Tokyo, Sanehiko Yamamoto. However, there are various accounts as to the precise details surrounding the invitation. Writing twelve years after the visit, Yamamoto recalled that Einstein's invitation had come amidst the "sudden ideological change" in Japan between 1919 and 1921, a notable sign of which was the publication by Kaizo-Sha of a novel by the Japanese Christian pacifist and labor activist Toyohiko Kagawa.[19] The novel's huge commercial success provided the funding for inviting Einstein and other prominent intellectuals, such as John Dewey, Bertrand Russell, and Margaret Sanger, to Japan. When asked to name the three greatest people in the world during his visit to Tokyo, Russell had allegedly told Yamamoto: "First Einstein, then Lenin. There is nobody else." Yamamoto wrote that he had then consulted the philosopher Kitaro Nishida and the science writer and former professor of physics Jun Ishiwara about relativity, in consequence of which he decided to allocate some 20,000 US dollars for Einstein's visit and dispatched his European correspondent Koshin Murobuse to visit Einstein in Berlin to discuss the terms.[20] However, according to Jun Ishiwara, Yamamoto had consulted with Nishida and himself some nine months prior to Russell's visit to Japan.[21] In yet a third account, Aizo Yokozeki, a former executive director of Kaizo-Sha, claimed that when Ishiwara suggested that the publishing house invite Einstein, they "did not know what to do." They consulted with Tokyo physicist Hantaro Nagaoka, who, while recommending that they invite Einstein, noted that the Japanese universities themselves had no funding to send Japanese students abroad or to invite Einstein to Japan.[22]

In any case, there seems to have been some lack of coordination concerning the initial invitation. In late September 1921, Ishiwara extended an invitation to Einstein on behalf of the *Kaizo* journal to visit Japan for a month-long tour in exchange for an honorarium of ¥10,000 (approximately £1,300).[23] At around the same time, Koshin Murobuse invited Einstein for a three months' visit, also on behalf of the *Kaizo*, and offered £2,000 for his lecture tour and travel expenses.[24] Even though, in a moment of fatigue earlier in the year, he had already exclaimed that he was "sick and tired of lecturing on

1. Einstein and Elsa with Sanehiko and Yoshi Yamamoto, Morikatsu and Tony Inagaki, Jun Ishiwara, and others, possibly Tokyo, 22 November 1922 (with permission of the Estate of Kenji Sugimoto and Kodansha Ltd. and courtesy of the Albert Einstein Archives).

relativity theory,"[25] Einstein responded positively. He thought of traveling to Japan in the fall of 1922. However, by early November, he had become displeased with the proposed financial terms: "The Japanese are real swindlers. They can get lost, for all I care. But I won't travel there under such lousy conditions."[26] Soon thereafter he canceled the visit altogether, citing the discrepancy in the terms of the two invitations as the reason for the reversal in his decision.[27]

However, merely five weeks later, Yamamoto, on behalf of the Kaizo-Sha, sent Einstein an official invitation and a contract with the following terms: a series of six scientific lectures in Tokyo and six popular lectures in various Japanese cities. The honorarium, which would include travel and lodging costs, would amount to £2,000.[28] In mid-March 1922, Einstein admitted to his close friend and colleague Paul Ehrenfest that he had been "unable to resist the sirens of East

Asia."[29] Shortly thereafter he wrote to Ishiwara that his planned departure for Japan would have to be delayed by one month due to his expected attendance at the annual conference of the Society of German Scientists and Physicians in Leipzig in September.[30] By August, the Imperial Academy of Japan had passed a resolution to welcome Einstein, and the Japanese government was preparing "a friendly welcome."[31] However, a letter from Uzumi Doi, one of Nagaoka's own students at Tokyo Imperial University, in which he expressed his strong criticism of relativity, indicated that not everyone would welcome Einstein with the same enthusiasm. Doi urged him "to return to the orthodoxy, freeing from the destructive spell that you have been fallen under so long since you were a youth of 26 years of age."[32]

During the negotiations regarding the invitation, Einstein revealed part of his motivation for his desire to embark on a trip to the Far East: "The invitation to Tokyo pleased me a great deal, as I have been interested in the people and culture of East Asia for a long time."[33] Three weeks after his arrival in Japan, Einstein wrote an article on his impressions so far and returned to the issue of his reasons for accepting the invitation: "when [Sanehiko] Yamamoto's invitation to Japan arrived, I immediately resolved to embark on such a great voyage that must demand months, even though I am unable to offer any excuse other than that I would never have been able to forgive myself for letting a chance to see Japan with my own eyes pass unheeded. [ ... ] in our country this land is shrouded more than any other in a veil of mystery."[34]

News of the upcoming tour in Japan reawakened previous plans to invite Einstein to lecture in China. However, the negotiations were complex and, to some extent, contentious. In September 1920, Yuanpei Cai, rector of Beijing University, had invited Einstein to lecture there.[35] Apparently, Jia-hua Zhu, a geologist from that university visiting at the University of Berlin, had conducted negotiations with Einstein in person on behalf of Cai. According to him, Einstein had promised that, after his tour of the United States, his next overseas visit would be to China. In March 1922, Zhu wrote to Einstein to renew the discussions after being informed of the planned tour in Japan. He told Einstein that his university actually wanted to invite him to stay for an entire year. However, he understood from the Chinese Embassy in Berlin that Einstein had informed them that, in light of his commitments in Japan, he only had time for a two-week lecture series in Beijing.[36] Einstein replied that, at the time, he had been unable to accept either the financial terms or the long duration initially proposed. However, circumstances had now changed, as the financial compensation from Japan enabled him to honor his prior commitment to visit China. He therefore agreed to a two-week visit.[37] In early April, Chenzu Wei, the Chinese envoy to Berlin, conveyed rector Cai's proposal for a course of lectures and a monthly honorarium of 1,000 Chinese dollars (about £120).[38] Einstein restated his willingness to lecture, but requested a significantly higher honorarium.[39] By late July, the Imperial University in Beijing had accepted his conditions.[40]

As with the foreseen visit to China, the genesis of Einstein's plan to tour Palestine and of the initial preparations for his arrival was also somewhat convoluted.[41] However, in contrast to the planned trips in the Far East, financial compensation was not an issue. Already in late 1921, in the wake of his somewhat successful trip to the United States, Einstein seems to have envisaged a visit to Palestine, where he could view for himself the settlement activities of the *Yishuv*, the local Jewish community. At the time, Chaim Weizmann had advised him that "there is as yet no great urgency to travel to Palestine." However, he did entreat Einstein to undertake another two-month tour of the United States.[42] The first evidence we have of Einstein's renewed interest in touring Palestine can be found in papers from late September 1922. After he had met with Einstein on the very day of his departure from Berlin for the Far East, the German Zionist leader Kurt Blumenfeld (who had played a crucial role in Einstein's initial contacts with the Zionists in Berlin) noted that Einstein had agreed to accept an invitation to visit Palestine by Arthur Ruppin, the director of the Zionist Organisation's Palestine Office in Jaffa.[43] He planned to travel to Palestine on his return trip from Java back to Berlin.[44] Yet this visit was to be a brief one—only ten days. According to Blumenfeld, Einstein emphasized that this short sojourn was "not to be confused with his actual trip to Palestine." He quoted Einstein as saying "one should only travel directly to Palestine and not incidentally following a trip to other countries."[45] Reports on the planned truncated trip appeared in the press on 6 October and the following days. It was immediately clarified that this was not to be the originally planned trip and that Einstein would remain in Palestine "for several months" after his return from the Far East. It was also reported that the goal of this brief trip was "to acquaint himself with conditions in the country" and he would "in particular [ . . . ] pay a visit to the Hebrew University in Jerusalem where it is believed it will be possible for him to deliver several lectures."[46]

However, correspondence regarding the trip between Einstein and Weizmann and among Zionist functionaries took place in parallel with these initial reports in the press. Apparently unaware of Blumenfeld's recent meeting with Einstein, Weizmann suggested to the latter that "perhaps you will also pass by Palestine and could on your *return trip* take a detour from Port Said." As Einstein had already departed from Berlin, his step-daughter (and secretary) Ilse Einstein replied to Weizmann that her parents would travel to Palestine unless they were prevented from doing so by extraordinary circumstances. In her opinion, it would be "a great joy for them" were Weizmann to also be there during their planned visit. She surmised they would arrive there in February. However, Weizmann had to visit Palestine urgently in November 1922 and subsequently embarked on an extended fundraising tour of the United States.[47]

Ten days following Weizmann's letter, Ruppin notified the Zionist Executive of information he had received from Blumenfeld. Einstein had accepted the invitation and planned to arrive in Palestine in late February or early March. Ruppin believed the visit may possess "great propaganda value" for the Zionist Organisation, and, in particular, for the project of the planned Hebrew University. As a consequence of the organization's experiences with Einstein during the U.S. tour (during which the illustrious guest had strayed from the "party line"),[48] Blumenfeld thought it was "absolutely necessary" not only that Einstein be accompanied by "Sicma [*sic*!] Ginzburg"[49] but that Elsa Einstein be accompanied by a woman, preferably by Rosa Ginzburg.[50] Thus, during this trip, both Einsteins were to receive official "shadows." Regarding the university, the Zionist Executive should contact Weizmann and find out from him "which projects and explanations you should give to Prof. Einstein."[51] Simultaneously, Ruppin asked Weizmann to tell the Zionist Executive "which opinions about the university project are to be viewed as the official stance." This was deemed necessary, as, according to Blumenfeld, if one does not give Einstein any directives "the danger exists that he endorse opinions presented to him randomly and this could diminish the propaganda effect of his tour."[52]

Einstein's contacts with the Spanish scientific community about a possible visit there began in 1920. At least two attempts to invite Einstein to Spain preceded his lecture tour in late February and early March 1923. In April 1920, the Argentinian mathematician Julio Rey Pastor had invited him to give a series of lectures in Madrid and Barcelona. Einstein informed his close colleague Fritz Haber that he "would [ ... ] definitely have to go to Spain."[53] However, the trip did not materialize. More than a year later, in July 1921, the mathematician Esteban Terradas é Illa invited Einstein to lecture at the University of Barcelona during the winter or spring semester. However, Einstein declined, hoping to visit in 1922–1923.[54]

Merely days before a dramatic turn of events, which would place his planned trip to the Far East in an entirely different context, Einstein hinted to his close friend Heinrich Zangger that in light of his recent trip to Paris and the turmoil surrounding his membership in the League of Nations' International Committee on Intellectual Cooperation, he longed for a change of scene. He confided that he "yearned for solitude," and the voyage to the Far East would mean "twelve weeks of peace on the open sea."[55]

Six days after his letter to Zangger, on 24 June, Germany's foreign minister, Walther Rathenau, was gunned down in the street in Berlin in broad daylight by right-wing extremists.[56] The assassination, which constituted a landmark event in the fledgling Weimar Republic, provoked outrage in most sectors of German public life, with the exception of the radical right. In its immediate aftermath, mass rallies and strikes were held, tensions ran high, and a possible civil war loomed. The day of Rathenau's funeral was marked by large demonstrations in support of the Republic. In a speech in the Reichstag, Chancellor Josef Wirth attacked the conservative enemies of the Republic as being complicit in the murder. The government issued various decrees against right-wing associations, and, eventually, legislation to protect the Republic was adopted.[57]

The assassination also represented a watershed moment in Einstein's life. Already painfully aware of the larger political implications, the assassination made him recognize that, as a prominent Jewish left-wing figure in German public life, he was in actual physical danger. In his letter of condolence to Rathenau's mother, Einstein lauded his friend, who would be recorded in the annals of history "not just as a man of great understanding and ability to lead, but also as one of the great Jewish figures to offer up their lives to the ethical ideal of reconciliation among peoples. [ . . . ] I myself feel the separation from him as an irreplaceable loss."[58] In a published memorial statement, he wrote that it "is easy to be an idealist when one is

2. Scene of Walther Rathenau's assassination, Berlin-Grunewald, 24 June 1922 (courtesy of Granger Collection).

living in Cloud-Cuckoo-Land; but [Rathenau] was an idealist even though he was living on Earth and knew its odor as others rarely do." But he was also candid in his criticism of Rathenau: "I regretted that he became a minister. Given the attitude of a large majority of the educated class in Germany toward the Jews, it is my conviction that proud reserve by Jews in public life would be the natural thing. Still, I would not have thought that hate, delusion, and ingratitude would go so far."[59]

Einstein's first wife, Mileva Einstein-Marić was "horrified" about the news that Einstein was "among those people whom certain elements—I don't know which— are plotting against."[60] Berlin journalist Friedrich Sternthal pleaded with Einstein to consider taking precautionary measures for his personal safety, warning of the "unbridled hatred" against him in "German-*völkisch* and similar circles."[61] Hermann

Anschütz-Kaempfe, a close collaborator and friend, invited Einstein for an extended stay in Kiel. While the invitation was not directly tied to the assassination, its timing—merely a day after the murder—was most likely not coincidental.[62] Einstein agreed to arrive with his wife a week later, writing: "Rathenau's assassination deeply shocked me and generally caused a great stir. It is unfortunately doubtful whether the Reich government will succeed in gaining mastery over all resisting elements. The army seems to be particularly unreliable. The old traditions of contempt for morality—fabricated for purposes of foreign policy—are now taking their toll inside the country." In his mind, the issues were not restricted to Berlin. He deplored the ongoing imprisonment of the prominent dramatist Ernst Toller in a Bavarian jail, and lamented: "Oh, nation of poets and thinkers, what has become of you!"[63]

The assassination now rekindled Einstein's earlier desire to leave Berlin for good.[64] On 11 July, he informed Marie Curie-Skłodowska that he would be resigning from the Prussian Academy of Sciences and from the directorship of the Kaiser Wilhelm Institute of Physics, thus removing himself from "clamorous Berlin in order to be able to work in peace again."[65] A day later, he informed his friend and colleague Max von Laue that, "officially," he was already away from Berlin, though he was physically still there.[66] He was now contemplating working for Anschütz-Kaempfe at his gyrocompass factory and purchasing a villa in Kiel.[67]

Einstein reversed himself a mere four days later, as the initial panic in the wake of the murder had subsided. Upon "calm reflection," he wrote that he would continue residing in Berlin, noting that there would not be much work for him at the factory in Kiel. As Elsa Einstein saw it, although Einstein had been strongly affected by Rathenau's murder and "he just had the feeling: get away from here to work in tranquility," by now "he realized that this thing with tranquility is an illusion. Nowhere else can he better duck out of sight than here in Berlin." Nevertheless, she confirmed that, "after the trip to Japan, he wants to give up his official offices here."[68]

The assassination affected Einstein's willingness to make public appearances. He decided not to deliver his planned address at the centenary celebration conference of the Society of German Scientists and Physicians, to be held in Leipzig in September. To his close colleague Max Planck, he wrote of the personal threats, having been warned against staying in Berlin, and against "making any kind of public appearances in Germany. For I am supposedly among the group of persons being targeted by nationalist assassins." In a draft of this letter, Einstein had pointed directly at "German-*völkisch*" elements who "were out to kill" him. He blamed the press for his predicament: "The whole difficulty arises from the fact that newspapers mentioned my name too often and thereby mobilized the riffraff against me."[69] In any case, the scene was set for Einstein's prolonged absence from Berlin.

## Analysis of the Travel Diary

In the following, we will try to "unpack" the deeper layers of Einstein's travel diary through a close reading of its text and other relevant historical sources.

We will pay special attention to Einstein's perceptions of the national and ethnic groups he encountered on his voyage and place these comments in the context of historical and cultural studies on Western images of the Orient and "the Oriental." We will examine the contemporary contexts of each of the countries Einstein visited. Einstein's preconceptions of these groups prior to his journey and the lens through which he perceived the regions he toured will also be explored. We will consider how he expressed his perceptions during the course of his travels and whether they changed due to his new encounters. Another topic will be to what extent the trip changed his own perceptions of himself—as a Jew, a German, and a European. In how far did the trip transform his concepts of the Self and the Other? What were his views on the notion of national character? A comparison of Einstein's perspectives with contemporary Orientalist, colonialist, and racist views will also ensue.

Furthermore, we will explore the impact Einstein's travels had on the countries he visited. What influence did his visits have on the societies that welcomed him? How did the local press react to his visit? How were popular perceptions of Einstein shaped by the masses' encounters with him? What political and diplomatic factors were influenced by his tours? How were the various scientific communities impacted, and what was their reception of relativity? What was Einstein's role as a disseminator of scientific knowledge?

Finally, we will consider what this trip meant for Einstein personally, examine the nature of his travel, the manner in which he traveled, and what personal impact the trip had on him. We will also explore the broader conclusions that can be drawn about Einstein's personal beliefs, individual prejudices, and ideological constructs.

### *Einstein on the Levant and Levantines*

The first region Einstein traveled through en route to the Far East was the Levant. It is very likely that, prior to his trip, Einstein's association with the word "Levantine" was not a positive one. In October 1919, Prague Zionist Hugo Bergmann, who was secretary of the Zionist Organisation's university committee at the time, expressed his concern about the academic standard of the planned Hebrew University in Jerusalem: "We do not want to do a shoddy job nor add a new university to the existing ones in the Levantine backwater but erect a good university that despite its limited resources still meets the full standard."[70] In a contemporary article, Bergmann articulated this opinion in an even more forceful manner: "[we] must be wary of erecting a university with a low *niveau*, a Levantine university, which will not add anything to our cultural richness and which no Jew will acknowledge as his national university."[71] Bergmann was apparently kicking down an open door with his advocacy of high academic standards for the Jewish university. A couple of weeks earlier, Einstein had expressed his support for such standards in a letter to fellow physicist Paul Epstein: it was "up to us to ensure that this university is on a par with its better-quality European sister institutions. There is certainly no lack of fine minds."[72] It seems fair to assume that Einstein did not have universities from "the Levantine backwater" in mind in his reference to "better-quality European sister institutions." One conclusion that we can definitely draw from both Bergmann's and Einstein's statements is that they defined their culture as decidedly European.

Einstein's perception of the Levant and Levantines underwent a significant change once he was confronted with the real thing. After five days at sea, the Einsteins reached Port Said at the northern end of the Suez Canal. Einstein's description of the scene on their arrival and their first encounter with the locals is extremely vivid and highly dramatic: "In the harbor, a swarm of rowing boats with screaming and gesticulating Levantines of every shade, who lunge at our ship. As if spewed from hell. Ear-deafening din. The upper deck transformed into a bazaar, but nobody purchases anything. Only a few handsome, athletic, young fortune-tellers are successful. Bandit-like filthy Levantines, handsome and graceful to look at."[73]

This passage reveals both the attraction and repulsion Einstein felt in his encounter with the Arab merchants at Port Said. What he is describing is an assault on almost all the senses. Even though the Europeans are on a Japanese ship, it is as if the local Orientals are attacking Western civilization itself. This is clear from the language Einstein employs: "swarm," "screaming," "lunge," "spewed from hell," "ear-deafening din," "bazaar," "bandit-like," and "filthy." Yet at the same time, the locals are attractive: "handsome and graceful to look at."

A hint at how Einstein perceives this scene unfolding before his eyes follows a couple of sentences later: "On the facing side [of the harbor] walls and buildings, one of those fierce colors which are often recorded on paintings of the tropics." Thus, his perception of the scene is greatly influenced by his visual (and also, as we shall see, textual) preconceptions of the Orient.

However, in accordance with the duality in Einstein's perception—attraction and repulsion—there are also well-behaved Levantines: by the time he has reached the port of Suez at the southern end of the Canal, he writes: "Arab minor merchants sail up. They are handsome sons of the desert, brawny, shining black eyes and better-mannered than in Port Said."[74] This description of the dark, proud Arab could be lifted from a nineteenth-century adventure story set in the Orient, say, by the famous German author Karl May.[75]

When Einstein arrived in Port Said a second time, on his return journey from the Far East, his reaction was no longer ambivalent, it only revealed a negative reaction: "City a real gathering place of foreigners with corresponding riff-raff."[76] The term "riff-raff" ("Gesindel" in German, which also has a connotation of "vermin") can well be interpreted as a xenophobic statement.

*Einstein on the Indians and the Sinhalese*

Available sources do not provide us with any insight as to Einstein's perception of Indians prior to his Far East voyage. After twenty days at sea, he encountered them in Colombo. When using a rickshaw, he was ashamed "for being complicit in such despicable treatment of human beings," yet felt helpless about it: "because these beggars in the form of kings descend in droves on any foreigner until he has surrendered to them. They know how to implore and to beg until one's heart is shaken up." This statement reveals his belief in a reversed nobility. In mentioning their existence on the city streets, he refs to their "primitive lives," which is fairly condescending. He also believes that "the climate prevents them from thinking backward or forward by more than a quarter of an hour," an attitude that reveals both Einstein's belief in geographical determinism and in the Indians' alleged intellectual inferiority. He notes laconically that the locals "live in great filth and considerable stench at ground level" and presumes to know that they have few needs: they "do little, and need little. The simple economic cycle of life." He opines that their cramped living conditions do not "allow any distinct existence for the individual." He compares them favorably with the loud Levantines he witnessed in Port Said: "no brutality, no market crying existence, but quiet, acquiescent drifting along, albeit not lacking in a certain lightheartedness." He also attributes the alleged stoicism of the Indians he encounters to geographic determinism: "wouldn't we too, in this climate, become like the Indians?"[77] This statement reveals his ambivalence toward the local population: he feels a certain degree of empathy for their endurance of their harsh plight, but he also seems critical of what they have become due to those hardships.

On their return journey, the Einsteins revisit Colombo. This time Einstein's description is fairly negative. He perceives the "natives" as "intrusive." One of their rickshaw drivers "was a completely naked primitive man." After a day trip, "the rickshaw coolies pounced on us" back in Colombo.[78] Here again, Einstein's European culture is symbolically being assailed. Furthermore, his use of the terms "coolies" and "primitive" to describe the locals is indicative of his feelings of superiority.

### Einstein on China and the Chinese

In contrast to the availability of sources in regard to Einstein's perception of Indians prior to his voyage, we do have a few instances in which he referred to the Chinese in the period preceding his trip. Intriguingly, in little more than a month, Einstein makes two statements about the Chinese, one of which is extremely positive, whereas the other is quite negative. In one of his first utterances on his emerging interest in Jewish nationalism and Zionist endeavors in Palestine, he remarked in March 1919: "I get most joy from the emergence of the Jewish state in Palestine. It does seem to me that our kinfolk really are more sympathetic (at least less brutal) than these horrid Europeans. Perhaps things can only improve if only the Chinese are left, who refer to all Europeans with the collective noun 'bandits.'"[79]

However, the following month, he writes to his Zurich friend Emil Zürcher about Russia being pillaged by "leaders of thieving gangs." According to Einstein, "these gangs are being mostly recruited from among the Chinese. Fine prospects for us, too!"[80] This could be interpreted as a fear that the Chinese could possibly take over Europe.

At the end of 1919, we find another revealing statement about the Chinese: "My friend [Michele] Besso is going back to the Patent Office. The poor fellow sets himself far too much apart from the animals—all conception and no will, the embodiment of Buddha's ideal. He would fit better in the East. This occurred to me particularly the night before last when I spent some time with a few refined Chinese; they have none of our obsession with purpose and practicality. Too bad for them and for the Chinese Wall!"[81] This is an interesting, rather ambivalent, statement. On one hand, Einstein seems to admire what he perceives as the positive effects of a Buddhist approach to life. On the other hand, he does not seem to believe that such an approach possesses any relevance for the West and, furthermore, he seems to imply that their lack of purpose and practicality would ultimately lead to the demise of their civilization.

A few weeks after his first encounter with Orientals in the Levant, Einstein reached Singapore, where he experienced an entirely different kind of Easterner.

Although he is mainly concerned with his attempts to raise funds for the Hebrew University in his interactions with Mannaseh Meyer, the leader of the Jewish community, during his brief sojourn in Singapore, Einstein does make some comments about the local Chinese inhabitants. Passing between two events held in his honor by Meyer, he notes: "then we drove through the Chinese quarter (tremendous bustle but there wasn't enough time to look, only to smell)."[82] Once again, Einstein's senses are being taxed by the indigenous population. The following day, he comments on his general impression of the Chinese. In his opinion, they "may well supplant every other nation through their diligence, frugality, and abundance of offspring. Singapore is almost completely in their hands. They enjoy great respect as merchants, far more than the Japanese, who are deemed unreliable."[83] Similar to his concerns in April 1919, Einstein seems preoccupied with the demographic implications of the Chinese birthrate. In addition, the phrase "almost completely in their hands" is also an indication of a perceived threat of domination.

A week later, he arrives in Hong Kong and again encounters the local Chinese inhabitants. His statements range from expressions of a considerable amount of concern for their plight to a certain degree of dehumanization. He first exhibits sympathy for the "stricken people, men and women, who beat stones daily and must heave them for 5 cents a day." In this way, "the Chinese are severely punished for their fecundity by the insensitive economic machine." In his mind, "they hardly notice it in their obtuseness, but it is sad to see." Thus, in spite of his feelings of sympathy, which seem similar to those one might express for an abused pet, he seems to deny the Chinese their full humanity. This is made even more blatant in the following passage when he visits the mainland: "Industrious, filthy, obtuse people. Houses very formulaic, balconies like beehive-cells, everything built close together and monotonous. Behind the harbor, nothing but eateries[,] in front of which Chinese don't sit on benches while eating but squat like Europeans do when they relieve themselves out in the leafy woods. All this occurs quietly and demurely. Even the children are spiritless and look obtuse." And then Einstein draws his racial (if not racist) conclusion from this scene: "It would be a pity if these Chinese supplant all other races. For the likes of us the mere thought is unspeakably dreary."[84] It seems clear that Einstein has bought —to some extent—into the perceived threat of "the yellow peril." Most intriguingly, it is a similar conclusion to the one he drew more than three years earlier in April 1919, yet exactly the opposite conclusion he had drawn merely a month prior to that, when he seems to have wished for the "horrid Europeans" to disappear and for only the Chinese to be left.

Throughout his stay in Hong Kong, Einstein notes the physical beauty, which he views as being in stark contrast to the dismal lodgings of the Chinese inhabitants. At the Peak, the highest point in Hong Kong, Einstein notes the "grandiose view." He mentions that the cable cars to the Peak segregated Europeans and Chinese, yet he does not express an opinion on such discrimination.

In an additional passage in the diary, Einstein quotes Portuguese teachers he meets who "claim that the Chinese are incapable of being trained to think logically and that they specifically have no talent for mathematics." He does not challenge this assertion in any way. To this xenophobia, he adds a healthy dose of extreme misogyny: "I noticed how little difference there is between men and women; I don't understand what kind of fatal attraction Chinese women possess which enthralls the corresponding men to such an extent that they are incapable of defending themselves against the formidable blessing of offspring."[85]

Einstein's next encounter with Chinese locals occurs in Shanghai a few days later. His harsh, dehumanizing comments continue: he finds a Chinese funeral "barbaric for our taste," the Chinese quarter with its "narrow streets" "is swarming with pedestrians," "in the air there is a stench of never-ending manifold variety." And he continues: "even those reduced to working like horses never give the impression of conscious suffering. A peculiar herd-like nation [ . . . ] often more like automatons than people." But there is a refuge from all this: when a Chinese meal does not agree with him, he seeks a "safe haven (to be understood literally)" with a German academic couple. When he visits a village in the countryside, he makes more dehumanizing assumptions about the local inhabitants: "We had a close look at the temple. The neighboring people seem to be indifferent toward its beauty."[86] As there is no way that Einstein could have known what the villagers were thinking, this was clearly a projection on his part.

On his return journey, he revisits Shanghai, where he reiterates most of the alleged national characteristics already mentioned. However, he also notes that "all are unanimous in praising the Chinese but also in regard to his intellectual inferiority in business skills."[87]

It is clear that Einstein's perception of the Chinese is full of contradictions: he feels both empathy for their miserable plight yet also dehumanizes them in a series of very disturbing statements. He also seems genuinely anxious about the possibility of their displacing all other peoples.

Studies on Western perceptions of China and the Chinese reveal to us that Einstein's impressions and stereotypes closely aligned with negative images of the Chinese that had been common in the West since the 1850s: they were perceived by many as the embodiment of darkness, barbarity, cruelty, and obscurantism and were characterized by dirt, stench, poverty, squalor, and superstition.[88] Just like Einstein's descriptions, Western perceptions noted the stark contrast between stunning scenery and squalid cities and villages.[89] However, intriguingly, right after World War I (thus shortly prior to Einstein's encounter with them), more positive images of the Chinese emerged in the West. The barbarity of the Great War had rendered some intellectuals more open to Eastern ideas.[90]

## The Aborted Tour of China

As discussed earlier, Einstein had intended to visit Beijing for two weeks and hold a lecture series there. However, even before his departure from Europe, political turmoil in China had weakened Einstein's resolve to follow through with his plans. He still held out the possibility that he would spend two to three weeks in China and lecture in Beijing and also possibly in some of the coastal cities.[91] Yet during Einstein's visit to Japan, his planned lecture tour in China ran aground due to serious miscommunication.[92] In early December, well into the Japan visit, Beijing rector Cai inquired about Einstein's expected day of arrival, enthusing that "the whole of China is ready to welcome you with open arms."[93] Two weeks later, Einstein replied that, "despite every good intention and despite earlier formal promises," it was now too late to accept an invitation. He stated that he had waited in vain for five weeks for a message from Beijing and had concluded they no longer wanted him to come. He expressed his hope that this "sad misunderstanding" could be redressed in the future.[94]

## Einstein on Japan and the Japanese

In contrast to Einstein's meager contacts with Chinese individuals prior to his voyage to the Far East, by the time he embarked on his trip, he had made some crucial connections with academics from Japan. Apparently his earliest Japanese contact was with the physics student Ayao Kuwaki in Bern in March 1909.[95] The following year, he praised an article on relativity dealing with ponderomotive forces by Japanese physicist Jun Ishiwara and stated that "this is [ . . . ] by far the only paper written on this topic that makes sense."[96] Ten years later, Einstein expressed that he was "extremely pleased" that Kuwaki (who had since become a professor of physics) had translated his popular book on relativity: "I still remember well your visit to Bern, especially since you were the first Japanese, indeed the first East Asian, whose acquaintance I ever made. You astounded me then with your great theoretical knowledge."[97] In the ensuing period leading up to his departure for the Far East, almost all his correspondence with Japanese individuals was related to the *Kaizo* invitation to Japan discussed above. However, recall that, at one stage during the negotiations, Einstein termed the Japanese "real swindlers" when he was dissatisfied with the financial terms of the invitation.[98] Soon afterward, he complained to Elsa that "now it looks like I'll be staying in Berlin for a very long time, because the curs[ed] Japanese put too much salt in my soup and watered it down as well."[99] Yet, as mentioned above, the hurdle of the compensation for the trip was eventually overcome.

Let us take a look at the concept of Japan Einstein formed prior to his visit. As we have seen, Einstein's yearning for the East formed part of his motivation for accepting the invitation to Japan. As historians of the West's perception of Japan have pointed out, this "lure of the East," was an important factor in travel to Japan.[100] This was increasingly the case following the "Exposition Universelle" in Paris in 1900, which had a great impact on making Japanese styles fashionable in the West. This craze for Japanese art and customs is known as "Japonism."[101] However, Einstein's interest in Japan was apparently also stimulated by even more exotic (and fanciful) sources. In interviews given during his tour, he mentioned the influence of the works of the Greek-Irish writer and journalist Lafcadio Hearn (who became a naturalized Japanese in 1896) on his perception of Japan prior to visiting there. Due to Hearn's writings from the "Lilliputian land," Einstein stated that, prior to his visit, he thought of Japan "in

3. Einstein and Elsa at Tokyo University of Commerce, 28 November 1922 (courtesy of the Hitotsubashi University, Tokyo).

fairy-tale terms of little houses and dwarfs."[102] In his article on his impressions of Japan, completed three weeks after his arrival, Einstein also pointed out the general sense of mystery he felt toward Japan: "Among us we see many Japanese, living a lonely existence, studying diligently, smiling in a friendly manner. No one can fathom the feelings concealed behind this guarded smile." In the article, Einstein admits that "all the things I knew about Japan could not give me a clear picture" of the country.[103]

The Japan Einstein toured in late 1922 was undergoing radical political, social, and cultural changes. The Taishō era, which had begun in 1912, was "a period when internationalism, cosmopolitanism, secularism, and democratization seemed to be replacing the parochial claims" of the previous Meiji era, when the country had primarily focused on nation building.[104] The post–World War I years saw the industrialization and modernization of Japan. Europe and America were starting to have an impact on Japanese culture. A class of intellectuals with an orientation toward the West was developing. Influenced by Christian, liberal, and radical ideas, the suffragist, labor, and student movements were emerging. Foreign scientists were extended invitations, and the first scientific periodicals appeared.[105] Even though the Taishō era saw the emergence of a spirt of liberalism, there was also an undercurrent of political terror: the year prior to Einstein's visit had seen the assassination of prime minister Hara Kei. Social welfare programs were also beginning to be developed. The democratic spirit was one of "imperial democracy": the foreign policy of the empire was supported by the mainstream parties—the distinction was not between pro-imperialist vs. anti-imperialist but rather "'slow-track' vs. 'fast-track' imperialists."[106]

What were Einstein's perceptions of Japan and the Japanese once he met them on board the steamer and after he arrived in the country, and how did his encounters transform his views?[107]

In almost breathtaking contrast to his harsh views on the Chinese, Einstein's perceptions of the Japanese could hardly have been more positive. He forms his first impressions of Japanese individuals on the first day of his sea voyage: "He (the Japanese) is unproblematic, [ ... ] he cheerfully fulfills the social function which befalls him without pretension, but proud of his community and nation [. . . .] He is impersonal but not actually withdrawn; for as a predominantly social being, he seems not to possess anything individually that he would have the need to be taciturn or secretive about."[108] Even though this is a positive description, Einstein seems to see the Japanese as less than fully human—he definitely perceives their individuality as not fully developed. His first recorded impression of Japanese women is almost caricature-like: he describes them as "crawling about [on deck] with children. They look ornate and bewildered, almost as if stylized. Black-eyed, black-haired, large-headed, scurrying."[109] After three weeks on board, Einstein does not seem to have penetrated the mysterious nature of his fellow Japanese passengers: "Japanese very devout. Weird fellows whose state is at the same time their religion."[110] His first encounters with Japanese music add to his sense of alienation: he finds their music "very foreign" and their singing made him "dizzy."[111]

Upon his arrival in Japan, he is immediately delighted with the country's landscapes. The first city he tours is Kyoto with its "magically lit streets." This impression is in stark contrast to his strong disdain for the Chinese cities he has just visited. He is also very taken by the "wonderful ancient Japanese architecture." The schoolchildren are "the dearest," the schools are "beautiful," and the vistas are "magnificent." And the "little houses and dwarfs" of Hearn's stories really do exist! He immediately notes the "cute little houses" and the "graceful little people" who "patter about, click-clack, on the streets." Einstein treasures the cleanliness, orderliness, and quiet manners he encounters in Japan. He finds the "Japanese unostentatious, decent, altogether very appealing," thus defining their national characteristics, as he perceives them.[112]

When he travels to the city of Nikko with Sanehiko Yamamoto, Morikatsu Inagaki, a staff member of Kaizo-Sha, and caricaturist Ippei Okamoto, he seems to really bond with them and to begin to perceive them as real individuals. When he has to deal with the German Society in Kobe, he admits that "at least in Japan, I much prefer dealing with Japanese."[113] He subsequently seems even more enchanted with his Japanese hosts.

However, in spite of all his admiration, following a conversation on the "Japanese worldview prior to contact with Europe," he is mystified by their alleged lack of scientific curiosity. His conclusion is far-reaching: "Intellectual needs of this nation seem to be weaker than their artistic ones—natural disposition?"[114] By asking himself whether this purported intellectual inferiority vis-à-vis the West is genetic, Einstein reveals that, in his opinion, there are differences in the intellectual capabilities of various nations.

After being in the country for three weeks, Einstein finally uses the word "cozy" ("gemütlich") to describe a get-together with his Japanese hosts.[115] Until then, "cozy" was a word that he used exclusively to describe his interactions with fellow Europeans, mostly Germans. It seems that by now he was humanizing the Japanese far more than during his initial encounters, when he described them in terms of mere caricatures. A few days later, he expresses his appreciation of his hosts' complete lack of "cynicism or even skepticism." And he then makes what amounts to a declaration of love to Japan and its people: "Pure souls as nowhere else among people. One has to love and admire this country."[116] And a week later, he writes to his sons that "the Japanese do appeal to me [ . . . ] better than all the peoples I've met up to now: quiet, modest, intelligent, appreciative of art, and considerate, nothing is for the sake of appearances, but rather everything is for the sake of substance."[117]

In his article on his impressions of Japan, Einstein points out some of the main differences he sees between Westerners and the Japanese. He highlights his perception that family cohesion is much stronger in Japan and states that the tradition of not expressing one's feelings enables the Japanese to live in close quarters, even by "those not in emotional harmony with one another." He rejects the notion that this leads to "inner impoverishment." He believes that even though he cannot delve into the Japanese mind, he can gain insight into their psyche through their art: "In this regard I hardly cease to be astonished and amazed. Nature and people seem to have united to bring forth a uniformity in style as nowhere else. Everything that truly originates from this country is delicate and joyful, not abstractly metaphysical but always quite closely connected with what is available in nature."[118] To a great extent, these comments by Einstein align with contemporary Western perceptions of Japan noted by historians. Accordingly, the essential qualities of Japanese art as conceived of by the West are "simplicity, functionality, minimalism." The Japanese are seen as being defined by their "love of nature" and their living "in harmony with nature." The Japanese garden is seen as the quintessential expression of "notions of harmony and a special Japanese affinity with the natural world."[119] Intriguingly, these perceptions ignore the reality that "many famous gardens [were] built during periods of war and social upheaval."[120] Furthermore, the Japanese actually have an ambivalent relationship toward nature.[121]

Einstein's encounter with Japanese women is one of polar opposites. At a dinner attended by "a number of geishas," he finds the dancing of the "very young" geishas and the "very expressive, sensual faces" of their older counterparts "unforgettable." In hinting at the "more unconstrained second part" of the dinner that would ensue after he and Elsa were "politely dismissed," Einstein implies that heavier drinking and interactions of a more flirtatious nature were about to take place. This is reinforced by his remark, which immediately follows, on his conversation with Inagaki on "geishas, morals, etc."[122] Whether Einstein adhered to the common Western misperception of geishas as prostitutes is unclear from his wording.

At the other end of the spectrum is his idealization of respectable Japanese women. By the end of his visit, Einstein perceives them as "flower-like creature[s],"[123] thereby revealing that his initial caricature-like perception of them has not advanced significantly beyond their external appearance. In his farewell letter to Yoshi Yamamoto, wife of Kaizo-Sha's president, he presents a highly idealized image of Japanese women and the Japanese family: "You will always epitomize for me the ideal form of Japanese femininity. Quiet, cheerful, [ . . . ] you are the soul of your home, which looks like a jewelry case, within which, like jewels, lie your darling little children. In you I rightly see the soul of your people and the embodiment of its ancient culture, directed primarily toward grace and beauty."[124]

4. Einstein and Elsa at a sake drinking party with geishas, Tokyo, possibly 25 November 1922 (courtesy of the Leo Baeck Institute, New York).

Thus, at one and the same time, Einstein perceives the respectable Japanese woman as an ornament and as the personification of the nation's soul. This can only make sense if what he means is that the ornamental nature of Japanese culture and society *is* its quintessential feature. To some extent, Einstein's polarized image of the Japanese woman fits into the antithetical Western stereotypes of the Oriental woman: the delicate, ornamental Madame Butterfly, and the Oriental femme fatale, the Dragon Lady.[125]

Einstein expresses his great admiration for Japanese art and sees it as a reflection of "the Japanese psyche."[126] He finds the Japanese theater "in part very exotic."[127] However, while he acknowledges his fascination with Japanese music, in the end he is quite critical of it and denies that it is "a major form of high art." He finds painting and woodcarving to be the highest expressions of Japanese art. What he admires the most about the Japanese artist is that "he loves clarity and the simple line above all else. A painting is strongly perceived as an integral whole."[128]

Einstein is ambivalent about Japan's adoption of Western culture. He accepts that "the Japanese rightfully admires the intellectual achievements of the West and immerses himself successfully and with great idealism in the sciences." However, he warns the Japanese not to "forget to keep pure the great attributes in which he is superior to the West—the artful shaping of life, modesty and unpretentiousness in his personal needs, and the purity and calm of the Japanese soul."[129] In this, Einstein was expressing a view similar to other Western visitors to Japan. Historians of travel to Japan have noted the reluctance of some Westerners to the westernization or modernization of the country "as it dilutes the 'otherness' of 'exotic' destinations."[130]

We can conclude that Einstein's perception of the Japanese underwent quite a radical change during his six weeks' stay in Japan. There was clearly a great disparity between his image of Japan prior to his trip and the reality he found once he boarded the ship, and even more so, after his arrival in the country. His view of Japan also evolved significantly during his tour. He gained greater insight into Japanese society and culture and seemed to increasingly humanize his hosts as his stay took its course. As with his views on the Chinese, many of Einstein's perceptions of the Japanese aligned with contemporary perceptions in the West: the admiration for Japanese art and architecture, the perceived unique affinity of the Japanese with nature, and the importance of the collective and the individual's subsummation in it. However, there were also less positive stereotypes of the Japanese in the West: they were perceived as being clannish, unfairly competitive, two-faced, and aggressive.[131] As we have seen, Einstein's stance was also ambivalent: he both admired them and was condescending toward them, especially with regard to their perceived inferior intellectual capabilities.

## Repercussions of the Trip Back in Germany

On 20 December, Einstein's peaceful ambling through temples and over mountains in Hiroshima Bay was abruptly interrupted by a telegram from the German ambassador, Wilhelm Solf, in Tokyo. Suddenly, his tour of Japan had led to political controversy in Germany. Solf informed his Foreign Ministry that, on 15 December, the *Japan Advertiser* had published a report according to which the German journalist and critic Maximilian Harden had testified at the trial in Berlin of his would-be assassins that "Professor Einstein went to Japan because he did not consider himself safe in Germany."[132] Solf feared this report might adversely impact "the extraordinarily beneficial effect of Einstein's visit for the German cause" and requested that Einstein allow him to deny this allegation by cable.[133] In his reply, Einstein admitted that a threat to his life had existed in the aftermath of Rathenau's assassination. While a "yearning for the Far East" had played a significant role in his accepting the invitation to Japan, the "need to get away for a while from the tense atmosphere in our homeland" had been a factor as well.[134]

## Einstein's Perception of Palestine

The Palestine that Einstein entered on 2 February 1923 was undergoing many changes on a variety of fronts. The British Mandate had been established in Palestine only two and a half years prior to Einstein's visit: Sir Herbert Samuel had assumed office as high commissioner in July 1920. Jewish immigration had intensified after World War I—this "Third *Aliyah*" (i.e., immigration wave) consisted mainly of newcomers from Eastern Europe. At the time of Einstein's visit, the Jewish population numbered 86,000 out of the general (predominantly Arab) population of approximately 600,000. Political tensions were high during this period: in early May 1921, attacks on

Jewish residents in Jaffa by Arab residents and the quelling of the riots by the Mandate authorities led to the deaths of 47 Jews and 48 Arabs. In June 1922, the Churchill White Paper was issued, which reaffirmed the Balfour Declaration, yet also declared that Jewish immigration would be restricted, as a reassurance to the Arab population. The Third *Aliyah* also led to a considerable increase in the number of agricultural settlements. In addition, foundations were laid for the industrialization and electrification of Palestine. In the three large cities, modern urban patterns began to crystallize. There were also significant achievements in the establishment of autonomous institutions of the *Yishuv*. The economic conditions in the country at the time of the visit were dire as a consequence of World War I. The Jewish economic section could not provide employment for all the Jewish population or for the new immigrants.[135]

Historians of travel to the Middle East have posed the question of whether visitors arrived in Palestine "already conditioned by the values of the society from which they [have] come" or whether they "accept[ed] local habits or customs without criticism."[136] Similarly, historians of German-Jewish travel to Palestine have stressed that travel accounts need to be seen against a "double background": "the social and cultural situation of German-speaking Jewry [and the] state of Palestine's Jewish population." Furthermore, the travelers visited the country with preconceived ideals, which they saw confirmed or contradicted by the reality they experienced, or, alternatively, they disregarded the reality and "project[ed] their ideal values onto the people of Palestine."[137]

Zionist travel to Palestine has also been viewed in the more general context of the cultural debate on Orientalism and the multiple (and often conflicting) perceptions of the "East." As Palestine was located in the Orient and was being settled predominantly by Eastern European immigrants, the Zionists had to grapple with the European perception of the Orient and Eastern Europe as "the twin counterpoints to the idea of a western, or European, civilization." For some Zionists, who believed in the "cultural mission" of the Europeans, the Orient was a "backward" region "into which western civilization must be imported." For others, Zionism's encounter with the East needed to be seen as fundamentally different from that of the European colonial powers' relations with the colonized. [138]

In light of these insights from historians of travel to the Middle East, we can pose the following questions: In what way did his ideological values and opinions impact how Einstein perceived the reality he encountered in Palestine, and were these values and opinions influenced by his visit in any way? What was the significance of the country for Einstein? Through what lens did he perceive the country? Did he perceive Palestine as a backward land into which European culture had to be imported, or did he subscribe to the views of those Zionists who believed that their efforts were fundamentally different from those of the European colonialists?

The initial entries on Palestine in Einstein's diary immediately reveal his sense of entering a land whose landscapes are very foreign to his central European eyes. When he first arrives in the country, he notes "a flat terrain with very sparse vegetation [ . . . ] olive trees, cactuses, orange trees."[139] While strolling through the Old City of Jerusalem, his most vivid impression was of both beauty and squalor. Einstein's perception of the traditional, pious Jewish community, the old *Yishuv*, which predated Zionist immigration to Palestine, was decidedly negative: "Then downwards to the temple wall (Wailing Wall), where obtuse ethnic brethren pray loudly, with their faces turned to the wall, bend their bodies to and fro in a swaying motion. Pitiful sight of people with a past but without a present."[140] His unfavorable perception of the ultra-Orthodox Jews may have been influenced by his childhood perceptions of unassimilated Jewry and possibly also by the Zionist perception of the old *Yishuv*.[141]

In stark contrast to his negative perception of the old *Yishuv*, his view of the modern Jewish community in Palestine, the new *Yishuv*, was extremely positive. He admired the dynamic entrepreneurial spirit of the new *Yishuv* and its urban development, the most salient expression of which was the rapid expansion of the first all-Jewish city, Tel Aviv.

One reason for Einstein's seemingly unbridled enthusiasm for the new *Yishuv* was possibly because internal conflicts in the community were hidden from his view by his Zionist hosts. Even though he toured private, cooperative, and collective farms, he does not seem to have been aware of the contemporary clash between private farm-owners in the rural farms (*moshavot*) and the pioneers (*halutzim*) on the *kibbutzim* over the issue of "Hebrew labor." This issue concerned the practice of the private farms hiring Arab laborers, whereas the *kibbutzim* wanted them to only employ Jewish workers.[142]

In stark contrast to this suppression of internal discord in the *Yishuv*, Einstein was well aware of the burgeoning national conflict between Jews and Arabs in Palestine. Einstein's visit came less than two years after violent riots in the Arab town of Jaffa. However, in all his accounts, he downplayed the potential explosiveness of the situation. Einstein's first impression of the Arab inhabitants of Palestine reveals his idealistic (and somewhat condescending) perception of them: "extraordinary enchantment of this severe, monumental landscape with its dark, elegant Arabian sons in their rags."[143]

During his tour, Einstein had limited direct contact with representatives of the Arab community. He only met with its moderate representatives: Jerusalem mayor Raghib al-Nashashibi, a few notables in the Galilee, and the Arab writer Asis Domet, who was quite a marginal figure in the local Arab population. Einstein seems to have apportioned equal blame to both Arabs and Jews for the nationalistic tensions in the region: "most of the difficulty comes from the intellectuals—and, at that, not from the Arab intellectuals alone."[144]

However, Einstein also possibly made some less positive remarks on the Arab population during his stay in Haifa. According to notes recorded by German Zionist Hermann Struck, Einstein stated that "if there were no Jews here, but rather only Arabs, the country would not need to export [produce], as the Arabs do not have any needs and live off whatever they grow themselves." If authentic, this disquieting statement would reveal a stereotypical perception of the local Arab inhabitants as having "no needs," which may have been informed by the mainstream Zionist image of the Arab populace.[145] We have already seen that Einstein subscribed to the colonialist view that the indigenous inhabitants of Colombo also had few needs. According to another utterance recorded by Struck, Einstein stated that he believed that the future of Palestine "will be ours" (i.e., Jewish).[146] While touring the north of Palestine, he expresses his sentiments against Arab landowners who "sold land to archaeologists at horrendous prices."[147] In this, Einstein seems to have been greatly influenced by the Zionist narrative that was being imparted to him on the tour.

Yet it was after his return to Germany that Einstein made his most disturbing comment on the conflict between Arabs and Jews in Palestine. He stated that, in comparison to "the two evils" faced by the agrarian settlers—debts and malaria, "the Arab question becomes as nothing."[148] That Einstein was prepared to implicitly refer to one party of the intercommunal conflict as "an evil" reveals the limits of his ethnic tolerance.

Einstein's diary reveals that he was enchanted with the landscapes and edifices of Palestine. However, in his later accounts, he makes it clear that it was his encounter with the *people* of Palestine which enthused him the most.[149] Of these, he particularly admired two groups: the young Jewish agrarian settlers and the Jewish urban construction workers. This followed from his support of the goals of labor Zionism and his hope that the Jewish people would play a more productive role than they had in the past from a social and economic point of view. The fact that the communal agrarian efforts were in the hands of young Russian-Jewish pioneers resonated with Einstein due to his special affection for young Eastern European Jews.[150]

In regard to Palestine's significance for Einstein, it is clear that immigration to the country, its colonialization, and its economic viability were all of secondary importance for him in comparison to its expected positive psychological impact on Diaspora Jewry. In his opinion, "Palestine will not solve the Jewish problem, but the revival of Palestine will mean the liberation and the revival of the soul of the Jewish people." He believed it would "become a moral center, but not be able to absorb a large section of the Jewish people."[151] Prior to his tour, Einstein viewed the land in instrumental terms, and this does not seem to have changed through his visit to the actual country. Intriguingly, once he returned to Berlin, there was no more discussion of his original plan to revisit the country for an extended stay in Palestine.

Similar to other tourists before him, the reality on the ground as he perceived it confirmed Einstein's views and opinions about the Zionist settlement efforts. At the same time, the achievements of the local Jewish community impressed him deeply, and he was confirmed in his belief that their efforts were a most worthy cause to which he would continue to devote a considerable amount of his time and energy. In particular, his positive perception of the (mostly) Russian settlers was confirmed, which increased his preexisting sympathy for young Jews from Eastern Europe.

Yet how can we categorize Einstein's perception of Palestine in the context of the historiography of travel to the region? Israeli historian Yeshoshua Ben-Arieh developed a typology of various perceptions by visitors to the Holy Land.[152] Einstein's perception fits three of these categories. To a limited extent, he perceived Palestine as "the land of the Bible and the holy places." However, this aspect was clearly not of great importance to him. To a considerable extent, he saw the country as an "exotic, oriental land"—this was especially the case in the Old City of Jerusalem and in his reaction to the desert landscapes and their inhabitants. Yet above all, Einstein perceived Palestine as "a land of new beginnings."[153] He believed in the positive effect of the modern urban developments in Tel Aviv and the ability of the agrarian colonization efforts to create "people of integrity,"[154] thereby reinforcing his sympathetic attitude toward these "new Jews."

How did Einstein perceive Palestine in relation to its location as part of the Levant? We can conclude the following. Even though he viewed Jerusalem's Old City as "Oriental-exotic,"[155] he seems to have perceived the new *Yishuv* (without explicitly saying so) as quintessentially European. Practically all the aspects he admired in the *Yishuv* were of European origin: the garden suburbs in Jerusalem, the faux-Oriental artwork at the Bezalel Art School, the "congenial Russians" on the *moshavim* and *kibbutzim*, the display of gymnastic exercises performed by the pupils of the "Herzliya" high school, and the modern facilities and factories in Tel Aviv and Haifa, to name just a few such aspects. Furthermore, as to the question of whether Einstein believed that Palestine needed to import European culture or blend in with its Levantine environment, he definitely advocated the Europeanization of the country. It is also apparent that Einstein's associations for processing the new visual information were European—he compared the local edifices he saw with similar European ones. Einstein clearly believed that Zionist efforts would be of great benefit to the peoples of the region. One can even see the focal point of his visit, his lecture on Mt. Scopus at the future site of the Hebrew University in Jerusalem, as the most blatant symbolic representation of this. Western knowledge was once again emanating from Zion.

In contrast to his perceptions of the other main countries he visited on his trip, there are no statements by Einstein that would reveal how he felt or thought about Spain or its people prior to his three-week sojourn there. However, we do have an indirect tidbit of evidence from one of his family members. In 1920, Einstein invited his step-daughter Ilse to accompany him on his planned trip to Spain. In response, she wrote that she was "constantly singing 'Beautiful Spain, far off to the south' in preparation for our trip."[156] Therefore, it seems safe to assume that Einstein perceived the country, to some extent, as a far-away, exotic land.

By the time Einstein reached Spain, he had been traveling for more than four months. Little wonder then that the diary entries for the Spanish leg of his journey are meager.[157] Consequently, we do not have much to go on with regard to his unmediated perceptions of the country and its people.[158] At the beginning of his stay in Barcelona, he refers to its "dear people." In Madrid, he comments on some of the individuals he meets. He finds the Nobel laureate Ramón y Cajal to be a "wonderful old thinker."[159] He describes King Alfonso XIII as "simple and dignified. I admire him for his manner." Of the Queen Mother, he writes: "The latter demonstrates her science. One notices that no one tells her what they think."[160] Einstein does not characterize the Spaniards as a collective in any way in his diary. The only snippet that provides a hint of how he perceived them was noted on the occasion of his receiving an honorary doctorate: "truly Spanish speeches with associated Bengali fire."[161] Thus, the title of the song Ilse Einstein sang to prepare herself for the visit she did not undertake was indeed indicative of Einstein's perception of the Spanish people as exotic.

Einstein was equally sparse with his comments on the leisure activities he partook of during his visit. He clearly relished both the cultural offerings and the day excursions. In Barcelona, he noted: "Folk songs, dances. Refectory. How nice it was!"[162] A painting by El Greco "is among the profoundest images I have ever seen." His entries on the excursions are terse but reveal his sincere enthusiasm: "One of the finest days of my life. Radiant skies. Toledo like a fairytale." And of El Escorial, he wrote: "Glorious day."[163]

### Impact of the Trip on Einstein's Perception of Europeans

How did Einstein's encounter with the inhabitants of the Orient impact his perception of his own reference group, the Europeans? We have seen that, in early 1919, a few months after the end of World War I, Einstein expressed his disgust at "these horrid Europeans."[164] Yet his perception of the Europeans he encounters during his trip oscillates between extremely positive and decisively negative poles. I have mentioned how he uses the term "gemütlich" ("cozy") almost exclusively for European friends and acquaintances. The home of his friends, the Pfisters, is a "safe haven" in Shanghai from the perceived Chinese onslaught on this Westerner's senses and abdomen.[165] Yet immediately following the end of his extended tour of Japan, he finds the Europeans in Shanghai to be "lazy, self-assured, and hollow."[166] However, following an admiring description of the stoicism of the most impoverished inhabitants of the Sinhalese quarter in Colombo, Einstein's disdain for Europeans is expressed in the strongest terms: "Once you take a proper look at these people [i.e., the Sinhalese], you can hardly take pleasure in the Europeans anymore because they are more effete and more brutal and look so much cruder and greedier—and therein unfortunately lies their practical superiority, their ability to take on grand things and carry them out."[167] All in all, Einstein clearly feels extremely ambivalent about the Europeans. In the diary, he expresses both attraction for and repulsion toward them.

Einstein visited the Middle and Far East during the age of imperialism. However, studies on Einstein have not mentioned how recent Germany's own colonial past was for him. For most of his youth and young adulthood, Germany was a colonial power, namely, from 1884 to 1919.[168] Furthermore, German colonialist ideology differed from its French and British counterparts in that the "Germans identified with the supposedly subjugated Other."[169]

During his visit to Hong Kong, Einstein clearly expresses his stance on colonial rule: "They [that is, the English] have an admirable understanding of governance. The policing is carried out by imported black Indians of tremendous stature, Chinese are never used. For the latter the English have established a proper university in order to bind those Chinese who have prospered closer to them. Who can emulate them in that? Poor Continental Europeans, you don't understand how to take the bite out of nationalistic opposition movements by means of tolerance."[170] This clearly reveals Einstein's enthusiasm for what he perceived as the "enlightened colonialism" and "civilizing mission" of the British. The method of coopting the local elites described here is well known from the history of colonialism. The colonized elite were trained by the colonial power to do their bidding in suppressing their own people.[171] In Colombo, too, Einstein voices his admiration for British colonial rule: "The English govern impeccably, without needless chicanery. I did not hear words of discontent against them from anyone."[172] It is interesting to note that during his visit to Palestine, Einstein does not express a positive or negative view of the British mandatory rule of the country. He does, however, note his admiration for what he perceives as the "lofty view of life" of the British high commissioner.[173]

Historians of colonialism have distinguished between two important terms—"projection" and "transaction"—to describe the relationships between the colonialist and the colonized. More traditional studies have focused on the colonialist's projection of their "desires, forebodings, or self-loathing" onto the colonized.[174] More recent studies have stressed the more reciprocal relations between the Europeans and the indigenous populations and have therefore preferred the term "transaction." The latter approach allows for the possibility of the exotic location having the potential to impact the identity of the colonialist.[175] This approach also speaks of "contact zones" between colonizer and colonized. These are "social spaces where disparate cultures meet, clash, and grapple with each other, often in highly asymmetrical relations of domination and subordination."[176] We can see Einstein's encounter with the rickshaw driver in Colombo as a crucial transaction in just such a contact zone. Even though he professes to be extremely reluctant to avail himself of the driver's services, in the end he resigns himself to the expectations of the situation and thereby collaborates in the colonialist oppression, albeit passively.

To contextualize Einstein's advocacy of "enlightened colonialism," it is crucial to point out that not all Europeans "were complicit to colonialism and imperialism to the same extent."[177] Some did not subscribe to a colonialist ideology at all. For example, the relationship of the early-nineteenth-century German explorer and geographer Alexander von Humboldt to the indigenous populations he visited and studied has been defined as "anti-colonial." Most crucially, von Humboldt acknowledged that, even if the indigenous are perceived as different, their basic humanity was never questioned.[178] However, as we have seen, there were times when Einstein denied the Other's humanity, at least to a certain extent.

### Einstein on the Orient and the Oriental

In his seminal work on Orientalism, the Palestinian-American intellectual Edward Said provides three definitions of the term. Only one of them is relevant for our purposes: "a Western style for dominating, restructuring, and having authority over the Orient."[179] Said's book has been hugely influential in colonial and cultural studies and has sparked a great deal of controversy and debate.[180] We cannot delve into that debate here, but the book's main point—namely, that the Orient in Western eyes was less a geographical reality and more of an ideological construct—is difficult to dispute.

Studies on Orientalism in general, and on German and Jewish Orientalisms in particular, are very instructive in helping us understand Einstein's image of the Orient. In the Orientalist ideology, the Islamic Orient is perceived as the "fundamental Other" for the Occident. The main purpose of Orientalism is to legitimize Western ideology and domination. The legitimization is accomplished by means of constructed binaries: the Oriental is seen as inferior, barbaric, wild, violent, uncivilized, childlike, irrational, fanatical, static, exotic, titillating, and (at times) sexual. In contrast, the West is perceived as superior, civilized, restrained, mature, rational, dynamic, enlightening, and familiar. Yet the dichotomies are actually more complex than that: the West is both attracted to and repulsed by the Orient; it perceives both sameness and difference, and displays both inferiority and superiority, toward the East. One of the reasons for the attraction and repulsion of the exotic Orient in Western eyes is that it represents the West's own "primitive" past.[181] Therefore, the origin of these images lies in their constituting projections of those qualities that the West rejects about itself.

German perceptions of the Orient and Oriental were greatly influenced by Karl May's works in the late nineteenth and early twentieth centuries. The author's alter ego, the protagonist Kara Ben Nemsi, is portrayed as the ideal German and European: his self-restraint is juxtaposed on the emotionality of the Oriental. However, May's perception of the Orientals is actually ambivalent: he sees them as both "full of temperament" and "hot-blooded." Accordingly, some Orientals can be enlightened by the West's civility, but others are unassimilable.[182] Einstein was thirteen when the first Kara Ben Nemsi novels were published in 1892, so he may well have read them and been impacted by them. We do not have any direct evidence of his having perused them.

Studies on Jewish Orientalism have examined the multiple layers of Jewish perceptions of the Orient: "Jews were deeply implicated in the twin conceptualizations of the Orient and of Europe's own orient at home. By the late nineteenth century, the notion of eastern Europe as *halbasien* [semi-Asia] [ . . . ] had acquired particular resonance in the context of the ever-present 'Jewish question.'" In the anti-Semitic discourse, "the Jew was a foreigner precisely because of his eastern origins and Oriental nature." For acculturated Jews, the Eastern European Jew "represented the 'Asian' element in Judaism which [they] had, happily, left behind." However, for the younger generation of Zionists, the *Ostjude* constituted a counter-myth of a more authentic Jew:[183] "a foil for the presentation of the western Jew as shallow, imitative, and assimilating."[184] In this Zionist perception, the Western Jew was seen as aged, weak, and ailing. In contrast, the Jewish pioneers in Palestine were healthy and authentic."[185] To most Zionists, the Orient was perceived as backward, a region into which Western civilization had to be imported.[186]

How does Einstein's image of the Orient cohere with these Orientalist ideologies?

When Einstein was a young lecturer in Bern, his sister Maja came to visit him. She asked a beadle which hall her brother was lecturing in and received an astounded reaction: "What, the . . . Ruski is your brother? And he was close to bestowing a far less flattering epithet on the Russian."[187] Thus, Einstein himself had been mistaken for an *Ostjude* and perceived as hailing from the European "Orient."

A few years prior to his actual encounter with the Middle and Far East, he saw himself and all Jews as Orientals. In August 1917, he justifies his rejection of the use of an express courier by referring to the Jews as "the sons and daughters of Asia," thereby implying that Asians do not do anything in haste. One day later, he advocates sloth as a virtue and recommends that his friend Zangger react with greater apathy to the vicissitudes of life and be more like "us lazy Orientals."[188] Even though these remarks were probably uttered tongue-in-cheek, they are still indicative of Einstein's self-image at the time. In the turbulent aftermath of World War I in Berlin, Einstein refers to the extreme uncertainty of everyday existence. He claims that, in reaction, people (including himself) have been "seized by a kind of Oriental fatalism, in which one feels good."[189] Einstein clearly seems to approve of this attitude of resignation. Thus, in contrast to the negative anti-Semitic perception of the Jews as Orientals, in all these cited instances, Einstein perceives being Oriental as a positive attribute.

The first mention of "the Orient" or "Oriental" in his travel diary occurs on the third day of his voyage as he is traveling through the Straits of Messina: "Stark, austere mountainous landscape on both sides. Towns likewise austere, predominance of the horizontal. Low, flat white houses. Overall impression: Oriental. Temperature rising relentlessly." There are only four other instances of the word "Oriental" used in the diary. In Singapore, the banquet held in his honor is held in "a spacious oriental refreshment hall." The "great humidity" in Singapore reminded Einstein "of a greenhouse. There is something orientally intoxicating about it." In Jerusalem's Old City, he describes the bustling scene: "Then diagonally through the (very dirty) city swarming with the most disparate assortment of holy men and races, noisy, and Oriental-exotic." And in Port Said, he refers to the governor as a "broad-faced Oriental."[190] Thus, he clearly associates stark landscapes, low-rise buildings, hot temperatures, high humidity, and a colorful mixture of races with the Orient. In contrast to his completely positive attitude toward the Orient prior to his trip, the instances cited in the diary are slightly positive, merely neutral, or, in the case of Jerusalem, rather negative. Intriguingly, he does not make any remarks about feeling like an Oriental during his trip.

Does Einstein subscribe to the Orientalist belief of European domination? As we have seen, he clearly advocated an "enlightened" form of colonialism. He definitely did not subscribe to any harsh treatment of the colonized, yet he did believe in the colonizing mission of the West. We can assume that his support can be explained largely by his mostly positive perception of the West's role in the modernization of the education of indigenous populations.

Does he apply the binaries of Orientalism to the local inhabitants he encounters? There are numerous instances in which he does, at least to some extent. For instance, he recorded his negative impressions of the uncivilized Levantines at the Suez Canal, the aggressive beggars in Colombo, the lethargic laborers and villagers in China, and the allegedly intellectually inferior Chinese and Japanese. Furthermore, titillating "voluptuous women" in Marseille[191] and flirtatious geishas in Tokyo both conform to stereotypes of the exotic as sometimes sexual. In these specific cases, one does not get the impression that Einstein perceived those instances of exoticism in a negative light. In many cases he also bestows explicitly positive attributes on the foreigners he meets (and some of them are ascribed to the same groups he attributes negative characteristics to, in alignment with the Western ambivalence to the Other). Some examples of these affirmative perceptions are the well-mannered Arabs at the Canal; the beggars in Colombo, who are noble and stoic; the Japanese, who are refined and cultured; and the Japanese women, who are delicate and gracious. He even has a substantial degree of sympathy for the harsh plight of the Chinese, whom he otherwise greatly maligns.

As for the influence of Jewish and Zionist perceptions of Orientalism, Einstein fully subscribed to the Zionist belief in the potential for spiritual rejuvenation of Western Jewry through the influence of the *Ostjuden*, especially of those who had settled in Palestine.

As for whether he felt like a European during the course of his journey, he seems to have experienced a great deal of ambivalence. We have seen that he occasionally expresses disgust with Europeans in the diary. Yet he definitely advocates the adoption of Western culture and science in the countries he visits, to a limited extent in Japan and, far more forcefully, in Palestine. In spite of the ambivalence, it seems fair to conclude that he feels very much like a European during his journey.

## Einstein's Gaze

Historians of modernism have identified "two powerful objectifying gazes—those of patriarchy, the [ … ] 'male gaze,' and of colonialism—the 'imperial gaze'"[192] In what way do these two gazes manifest themselves in Einstein's travel diary?

One crucial aspect of Einstein's gaze as a Western traveler are the preconceptions that influence the way in which he perceives the landscapes and individuals he encounters. The diary reveals that Einstein's perceptions are often filtered through European—in particular, Swiss and German—lenses. The mountains surrounding the bays in Hong Kong remind him of the Alpine foothills. On arrival in Japan in the straits of Kobe, his first impressions evoke associations of fjord-like imagery. A festively lit street in Kyoto is reminiscent of the *Oktoberfest*, and temples in the surrounding landscape remind him of Italian Renaissance architecture. In Palestine, the Sea of Galilee prompts comparisons with Lake Geneva. And in Nazareth, the German Inn feels like home.

The second aspect of Einstein's gaze is the actual act of gazing (or staring) that occurs in his interactions with the indigenous populations. Einstein first comments on this during the Einsteins' first visit to Shanghai: "With European visitors like us, comical mutual staring—Else particularly imposing with [her] seemingly aggressive lorgnette." The fact that the staring is mutual is indicative of the reciprocal relations between the Westerner and the local inhabitants. When they visit a village in the Shanghai environs where Europeans do not usually visit there takes place "reciprocal staring even more comical than in the city."[193] On their return journey from Japan, in the small town of Negombo near Colombo, Einstein expresses the mutual staring in an amusing, tongue-in-cheek manner: "We were ogled at everywhere just as the Sinhalese are back home."[194] His close interactions with indigenous locals heightens Einstein's awareness of his own identity as a Westerner. As E. Ann Kaplan has stated, "people's identities when they are traveling are often more self-consciously *national* than when they stay home."[195] Thus, one important consequence of his interactions with locals in the East is the intensification of Einstein's identity as a European during this trip. To some extent, the trip renders him even more European.

A further aspect of Einstein's gaze is its masculine quality. In the field of culture studies, the "male gaze" is a phrase used to describe the presentation of women as objects of male pleasure.[196] In numerous instances, Einstein makes a special note in his diary of the women he encounters. His journal thereby offers us several glimpses of Einstein's perceptions of women. His gazing at women can be divided into pleasurable experiences and those that were less so. In his very first entry in the diary, he notices the abovementioned "voluptuous women" in Marseille, prior to his boarding the ship. This is perhaps an indication of Einstein's desire at the very outset of his voyage for a more carefree existence on board the ship.

We have already seen how he perceived and related to Chinese women during his visits to cities with large Chinese populations. His descriptions of these women were mostly negative, viewing them either as victims or even denying them their femininity altogether.

And we have also noted his polarized view of Japanese women. On his return journey, he notes several instances of pleasurable experiences with women. Immediately after the end of his tour of Japan, when he arrives back in Shanghai, he sat "next to [a] fine Viennese lady" at a New Year's Eve celebration. This seems to have been the only high point of the evening: "otherwise noisy and, for me, doleful."[197] In Penang, he sees a "beautiful, intrusive beggar woman."[198] In Colombo, he is clearly impressed by "a ravishingly beautiful, fine young Sinhalese woman."[199] And in Palestine, he immediately expresses his pleasure in meeting Sir Herbert Samuel's "capable and earthy cheerful daughter-in-law."[200] She is the only woman—besides Elsa, of course—whom he refers to by her first name (Hadassah), during the entire course of the diary.[201]

Yet Einstein's gaze is not simply that of the "seeing [white] male [ . . . ] whose imperial eyes passively look out and possess,"[202] even though that is also an integral part of his gaze. As a German Jew and an insider/outsider, his was a "double-consciousness," which is the "sense of always looking at one's self through the eyes of others."[203] Thus, Einstein is also gazing at himself while being gazed at by others. During his trip, Einstein is caught between the "imperial gaze," in which he is the objectifier, and the non-Jews' gaze on the Jew, in which he is the objectified. And as a celebrity, he is even more caught in the gaze of others, and—in the course of his overseas travels—he was becoming even more objectified himself as he was emerging as an international cultural icon.

### Einstein and the Other

Cultural studies on alterity can provide us with some insights into the representation of the Other in Einstein's diary. At the basis of the relationship of the traveler and the indigenous populations is the Self/Other dyad. Internally, the traveler projects a reflection of the

Self onto the Other.[204] The Other is "a canvas upon which the best and worst qualities of the self can be projected and examined."[205] The foreigners are, in Vamik D. Volkan's phrase, "suitable targets for externalization." What is too painful to deal with inside is projected onto the Other.[206] The Western traveler defines the foreign Other "not in terms of his or her own reality, but in terms of norms promulgated by the Self." The traveler uses a filtering lens to compare "resemblances and differences between his own known, familiar world and the unknown world of the Other."[207] In the course of this process, "the people to be othered are homogenized into a collective 'they,' which is distilled even further into an iconic 'he' (the standard adult male specimen)."[208]

How does the Self/Other dyad express itself in Einstein's journal? What qualities does he project onto the Other? We can conclude from his initial reading material at the outset of his voyage that the trip almost immediately evokes Einstein's desire for greater introspection. Thus, his focus on the Self is facilitated by his perusal of a treatise on physique and personality traits.[209] It is possible that his urge for greater self-understanding was prompted by both the expected leisure time on board and his anticipation prior to his trip of his encounters with the representatives of multiple nations. As for the filtering lens Einstein employs in his perception of the Other, we have seen that it was largely European in nature. In many instances in the diary, Einstein describes his interactions with the foreigners he meets as some form of assault. These can take the form of an attack on his senses (e.g., a cacophony of noise in the Suez Canal, or the stench in the Chinese quarter of Singapore). Alternatively, the attacks are of intrusiveness (e.g., the unrelenting insistence of the rickshaw drivers in Colombo, or the pleading of the Zionists for him to migrate to Palestine).

As for the qualities he disdains in foreigners, they include aggressiveness, pushiness, lethargy, backwardness, intellectual inferiority, and obtuseness. However, there is an interesting contradiction here: we have seen that, occasionally, Einstein advocates adherence to traditions, indifference, and laziness. The qualities in the Other that he admires are authenticity, cleanliness, orderliness, refinement, high-mindedness, and gracefulness. Contradictorily, he also, at times, likes earthiness, messiness, disorderliness, and grime.

Einstein employs the homogenization of the Other in two different ways. For example, he often writes of "the Japanese" in the singular to refer to the country's entire population. In contrast, he always (with one exception) refers to the Chinese in the plural. This seems like an expression of his greater emotional distance from the Chinese. This distance may be understandable, as he has far more time to get to know the Japanese. Furthermore, as we have seen, during the course of his tour of Japan, he increasingly humanizes his various hosts and companions as his relationships with them evolve.

### Einstein and National Character

A common way to differentiate between Self and Other is to stereotype the members of other nations and endow them with an immutable "national character." These national characterizations "function as commonplaces—utterances that have obtained a ring of familiarity through frequent reiteration." For the believer in national character, the familiarity and recognition value of these stereotypes are of greater importance than their empirical truth value. There are two layers to the use of national character. On the surface level, "the discourse of national stereotyping deals primarily in psychologisms, ascribing to nationalities specific personality traits." Yet on a deeper level, "a nation's 'character' [ . . . ] is that essential, central set of temperamental attributes that distinguishes the nation as such from others and that motivates and explains the specificity of its presence and behavior in the world." In the process of ascribing these essential qualities, "certain traits are singled out and foregrounded because they are typical" in two senses: "they are held to be representative of the type, and they are unusual and remarkable."[210]

From our detailed analysis of his perceptions of the foreign nations and ethnic groups he encounters on his trip, it seems abundantly clear that Einstein was a very firm believer in 'national character' and its essentialist qualities. He repeatedly employs generalizations and constructs stereotypes to convey to the reader his impressions of the members of the national groups with whom he interacts. We have also seen that he is very quick to do so. He forms his opinion on "the Japanese" (singular) on the very first day on board the ocean liner. He definitely believes that his initial impressions of a very small number of representatives of each nation grant him an insight into the alleged "character" of an entire national group. Repeated encounters only reinforce his first impressions.

Einstein's belief in "national character" is particularly intriguing for two reasons. First, as I have argued elsewhere, he was not a nationalist. In my study on Einstein and Zionism, I concluded that he was not a Jewish nationalist but rather a Jewish "ethnicist," that is, someone with "a state of mind of positively valued self-awareness" with regard to the ethnic attributes of his or her ethnic group, but, in contrast to a nationalist, does not subscribe to the need for political independence in their own ethnic territory.[211] Second, as he was a firm believer in empirical evidence in his science, his willingness to forgo such proof in his use of national stereotypes is quite fascinating. Thus, it seems fair to conclude that his penchant for national characterization was founded on deeper, emotional needs of his personality.

### Einstein, Race, and Racism

Both historical studies on the concept of race and sociological studies on racism can help us understand the ways in which Einstein uses the term "race" and whether his negative assessments of foreign nations and ethnic groups can be defined as racist.

In spite of their radical ideas, many prominent thinkers during the age of Enlightenment and in the first half of the nineteenth century, such as Voltaire, Kant, Hegel, and Marx, did not believe in the equality of the races.[212] In the late nineteenth century, the concept of race evolved as a "scientific paradigm" among significant sectors of the European intelligentsia, leading to the emergence of "scientific racism."[213] By the 1880s, the radical right-wing adherents of "biological anti-Semitism" raised claims of the alleged "hereditary defects" of the Jews with increasing frequency.[214] These developments led to the search by Jewish intellectuals for what they considered to be "objective criteria for defining Jewish national identity." Thus, racial ideas were introduced in the intellectual discourse of Jewish identity in Central Europe.[215] In Germany, it was mainly the Zionists who utilized "essentialist views" on Jewish identity to strengthen their arguments and to "instill national pride."[216] However, some moderate and extreme Jewish assimilationists also dabbled in racial definitions of Jewish identity.[217] Some of them, such as Otto Weininger, Arthur Trebitsch, and even Walther Rathenau, actually accepted the main arguments of "Aryan racial ideology."[218] Even some mainstream Zionist ideologues, such as Max Nordau, adopted the anti-Semitic claim of the alleged "degenerate qualities" of the Jews.[219] The main intra-Jewish reasons the concept of race emerged during this period as so important for defining a collective Jewish identity were a marked decrease in the relevance of religion among Western Jewish communities, their lack of a uniform language, and the absence of a self-contained area of Jewish settlement. Therefore, a common ancestry, a joint "biological" or "racial" heritage, based on supposedly "scientific" theories, came to be perceived as the foremost defining characteristic of Jewish identity.[220] Many German Zionists advocated the Jews' "racial purity" and believed it needed to be "protected" in light of the increasing rate of intermarriage among the German Jewish community. Even moderate leaders like Arthur Ruppin, who supported left-wing Zionist goals and a peaceful coexistence with the Arabs in Palestine, subscribed to the ideal of Jewish "racial purity."[221] Some Zionists in Germany also espoused the "racial otherness" of the Jews and concluded that the Jews and the Germans were incompatible entities.[222] However, historians of German Jewry have maintained that the proponents of racial theories among Jewish nationalists "never explicitly embraced the doctrine of racial superiority."[223] At the same time, this did not prevent German Zionist propaganda from "assertions of the Jews' alleged moral, intellectual, and emotional advancement."[224]

Recent sociologists of racism have defined the phenomenon at its most basic level as "a denial of humanity [ ... ] and a means of legitimating inequality."[225] The dialectic of the Self and the Other "is found at the core of all racisms."[226] The process of differentiating one's own group from the Other (i.e., "racial categorization") can be defined as "a process of delineation of group boundaries and of allocation of persons within those boundaries by primary reference to (supposedly) inherent and/or biological (usually phenotypical) characteristics."[227] Miles and Brown define racism as an ideology in which "biological and/or somatic characteristic(s)" are used "as the criterion by which populations are identified. [ ... ] these populations are represented as having a natural, unchanging origin and status, and therefore as being inherently different." The targeted group "must be attributed with additional (negatively evaluated) characteristics [ ... ] Those characteristics or consequences may be either biological or cultural."[228] The negatively evaluated group "is represented ideologically as a threat." The ideology "may take the form of a relatively coherent theory." But it can also appear "in the form of a less coherent assembly of stereotypes, images, attributions, and explanations that are constructed and employed to negotiate everyday life."[229]

How does Einstein employ the term "race" prior to his trip to the Far East? We find the first instance of his use of the concept during World War I. In defining the term "nationality," he names race, community, language, and "possibly religion" as its notable components.[230] About two years later, he makes a similar point, this time listing race, temperament, and traditions as the essential qualities of Jewish nationality. In an attempt to explain anti-Semitism, Einstein claims that hatred of the Jews is not caused by their idiosyncratic characteristics but rather by their very existence. For him, this disdain toward "the members of a foreign race" is inevitable.[231] Two days later, in a caustic letter to the central organization of assimilationist German Jews, he even asserts that anti-Semitism is not a harmful phenomenon, as it may be the reason "that we have survived as a race."[232] Over a year later, his view of the role of race in the formation of Jewish nationality seems to have shifted: "I have not reached a decisive opinion on the extent to which we Jews should regard ourselves as a race or a nation or on the extent to which we form a social community solely out of tradition."[233] We cannot be certain what caused this change in Einstein's perception, but it may somehow be related to his encounter a few months previously with the leadership of the Jewish community in the United States—with whom he was not enchanted—and with the Jewish masses in America—who turned out to give him an enthusiastic welcome and whom he perceived of quite positively.[234]

Einstein uses the term "race/s" three times in his Far East travel diary. After encountering Jews of Middle Eastern descent in Hong Kong, he notes: "I am now convinced that the Jewish race has maintained its purity in the last 1,500 years, as the Jews from the lands of the Euphrates and the Tigris are very similar to ours." That same day, he makes the disturbing statement we have already analyzed about how it "would be a pity if these Chinese supplant all other races."[235] And, finally, he describes the scene in Jerusalem's "very dirty" Old City "swarming with the most disparate assortment of holy men and races, noisy, and Oriental-exotic."[236]

What can we conclude from all these sources about Einstein's perception of race, and can he be viewed as a racist? In his earlier statements, it is clear that, when he uses the term "race," he is referring to a common ethnic origin that is biological in nature. The Jews are a unique entity, almost its own organism, which has "survived." It is evident that he—similar to other German Jewish intellectuals of his time—is using racial categorization as one way to define Jewish national identity. He definitely views the Jews as a "race" who are easily distinguishable from other races and peoples. We know that he ascribed phenotypical characteristics as a means for Jews and non-Jews to differentiate themselves from each other: at elementary school, he was struck by "the children's remarkable awareness of racial characteristics."[237] As he does not ascribe any superior characteristics to the Jews, it would seem that at this stage, while his views are racialist, they are not racist.

In the context of Einstein's belief in phenotypical characteristics as ethnic markers, we should mention his handwritten caption to Ippei Okamoto's famous caricature of him with its exaggerated, large nose, which can definitely be seen as a stereotypically "Jewish" nose. The caption reads "Albert Einstein or the nose as a reservoir for thoughts."[238] The inscription is both self-mocking with regard to his own "typical" Jewish nose and mocking of the widely held anti-Semitic stereotype of a large nose as clear sign of alleged Jewish ethnicity. The ironic implication of the caption is that there is an obvious correlation between the Jews' characteristically large noses and their pronounced intellectual abilities. But, as we have seen, Einstein subscribed to such stereotypes himself, and, as they were acquired during his early childhood, it seems safe to assume that they were part of a deeply ingrained belief.

The ways in which the word "race" is used in the diary adds to our understanding of Einstein's concept of the term. Suddenly, the Jewish race is something that has "maintained its purity," thereby implying that, potentially, it could also be tainted by what he may have perceived as miscegenation. This statement reveals that Einstein seems to have agreed with those German Zionists, such as Arthur Ruppin, who deemed that the "racial purity" of the Jews needed to be protected. The description of the bustling scene in Jerusalem's Old City with its use of words with negative connotations (e.g., "dirty," "swarming," and "noisy") betrays a subtle racist intention. In contrast, Einstein's diary entries on the biological origin of the alleged intellectual inferiority of the Japanese, Chinese, and Indians are definitely not understated and can be viewed as racist—in these instances, other peoples are portrayed as being biologically inferior, a clear hallmark of racism. The disquieting comment that the Chinese may "supplant all other races" is also most revealing in this regard. Here, Einstein perceives of a foreign "race" as a threat, which, as mentioned above, is one of the characteristics of a racist ideology. Yet the remark that must strike the modern reader as most offensive is his feigning to lack an understanding for how Chinese men can find their women sufficiently attractive to have offspring with them.

In light of these instances, we must conclude that Einstein did make quite a few racist and dehumanizing comments in the diary, some of which were extremely unpleasant, especially for the modern reader. In these private remarks, Einstein espoused the alleged inferiority of other "races." In doing so, he seems to have surpassed the public "racialism" of German Jewish and Zionist intellectuals mentioned above, who, according to historical studies, did not claim any superiority of the Jewish people.[239] I certainly do not believe that Einstein subscribed to a "relatively coherent theory" of racism. And he definitely did not actively advocate that any concrete measures be instituted against the alleged "threat" of other races. But he does not seem to have been greatly disturbed by some exclusionary measures, such as racial segregation in Hong Kong. Thus, even though he did not subscribe to a fully blown racist ideology, I would definitely maintain that, at times, Einstein's racism would fall under the category of "a less coherent assembly of stereotypes, images, attributions, and explanations"[240] that he used to justify the differences he perceived between the members of various ethnic groups.

## The Nature of Einstein's Travel

How can the history of travel and tourism in the modern era instruct us as to the nature of Einstein's travel during his trip to the Far East and beyond?

Travel is often seen as a journey into the self.[241] Tourism in the industrial age has also been defined as "the exceptional normal" by Italian microhistorian Edoardo Grendi.[242] The historian of tourism Dean MacCannell makes a similar claim when he writes that "tourists search for experiences, objects and places that enable them to recover structures from which they are alienated in daily life." Modern travelers search for greater authenticity in the foreign lands they visit.[243] In the age of colonialism, this entailed either "mentally processing the foreign in ways that often encompass it within the comforting categories of the familiar . . . or else constructing it as Other: exotic, erotic, inferior, superior, dangerous."[244] The quest for the "authentic Other" has also been seen as a quest for the "authentic Self."[245] The image of the foreigner is preshaped by conceptions of one's own culture.[246]

Many travel historians have distinguished between the "traveler," who is "trailblazing and self-motivated" and the "tourist," who is "reactive, following established channels and seeking prescribed experiences in predetermined ways."[247] Einstein's overseas trips took place against a background of the rise of modern mass tourism. In Germany, there was a sharp increase in the desire to travel following World War I. In contrast to the prewar period, travel was no longer seen as the privilege of the few but as a commodity accessible to more and more sections of the population.[248]

Can Einstein's trip to the Far East be viewed as a journey of self-exploration? Did it take place in an atmosphere of "the exceptional normal"? Was he trying to recover structures from which he was alienated?

From the second day of his voyage, Einstein seems to be set to use the trip as an opportunity for greater self-discovery and introspection. The rising temperatures seem to make this more feasible: "The sun revitalizes me and removes the gulf between 'ego' and 'id.'" The very next thing Einstein mentions is the fact that during his first days on board the ship, he is using his leisure time to read the abovementioned treatise on the connection between physiology and personality types by German psychiatrist Ernst Kretschmer. At first he does not find the book that relevant: "I can [ ... ] categorize many of my fellow beings but not myself, because I'm a hopeless hybrid." Yet the next day, he reveals that the book has had a profound impact on him, providing him with a significant insight into his own character. "Yesterday, was still unsettled by reading Kretschmer. Felt as if grabbed by a vice. Hypersensitivity transformed into indifference. During adolescence, inwardly inhibited and unworldly. Glass pane between subject and other people. Unmotivated mistrust. Substitute paper world. Ascetic impulses."[249]

Actually, on a symbolic and psychological level, one can interpret the trip as a descent into Einstein's psyche, a journey into his "Heart of Darkness." This is especially the case during the first few weeks of the voyage, when the temperature is rising dramatically and he is coping with tropical heat for the first time as he nears the Equator. His diary entries contain references to the "hothouse temperatures" and his being in a "vegetative state," and we have already encountered his reference to the Levantines being "spewed from hell."

What kind of a traveler was Einstein? There is no evidence that he ever consulted guidebooks or maps during any part of his entire voyage. To some extent, his tours were similar in nature to tourist package deals—naturally, not similar to mass package deals, of course, but rather package deals tailored for an individual, a VIP. This mode of travel was distinctly different from Einstein's past intra-European travel, where he was far more autonomous in his choice of destinations and itineraries. Einstein also used travel for frequent getaways from Berlin in places like Leyden, Kiel, Lautrach (in southern Germany), and Zurich. His overseas trips became a new way for him to stay away from "nerve-racking" Berlin for even longer periods.

What was the nature of Einstein's travel in Japan? His itinerary and schedule were very tightly managed by the Kaizo-Sha publishing house and the academic institutions that hosted him. The Einsteins only made two independent forays by themselves without local escorts. Even though this is understandable in light of the linguistic impediments, it is also indicative of the manner in which Einstein traveled. There is also no evidence that he was involved in choosing the itinerary of the Japanese lecture tour.

He seems to have been more than happy to be completely passive in choosing the destinations and practically left almost all decisions to others.

In addition to the distinction between the traveler and the tourist, historians of travel to the Middle East have added a third type of visitor, the pilgrim. The goal of pilgrims was to establish "a link between earth and heaven"; travelers focused on their own individual experience of the land, whereas tourists visited en masse in organized tours.[250] Was Einstein a pilgrim, traveler, or tourist in Palestine?

In regard to this question, we can conclude that the narrative, as reflected in his travel diary, was entirely secular in nature—the few allusions to Biblical figures do not render his trip to the Holy Land a religious pilgrimage. But was it a secular pilgrimage or was he a traveler or a tourist? Even though he insisted on traveling second class on the train to Lod (in spite of the *wagon-lit* that had been reserved for him),[251] there are no indications that Einstein endured any hardships during his visit. All his lodgings throughout his tour, from Government House in Jerusalem to private board in Haifa and the German Inn in Nazareth, were comfortable quarters. Also, nothing indicates that he wanted to establish "a link between heaven and earth" through his trip. The only manner in which we can define his visit as a secular pilgrimage is that he was looking for confirmation of his beliefs formed prior to his visit about the Jewish colonization efforts.

Can we, therefore, define Einstein as a "traveler" to Palestine? His tour of the country does not seem sufficiently independent to qualify as such. There are certainly no indications that he made any autonomous arrangements during his visit. It also does not seem that Einstein had the time (or the inclination) to prepare himself for the upcoming trip. The only extant book on Palestine in Einstein's personal library was sent to him by its author, the left-wing travel writer Arthur Holitscher, in August 1922. However, the book was never opened by Einstein.[252]

The sole aspect of his tour that would fit the definition of a traveler were Einstein's independent-minded views of the people and places he encountered. In spite of his great enthusiasm for the achievements presented to him by his Zionist hosts and his strong commitment to assist them in furthering their goals, we cannot conclude that he became entirely enthralled with their efforts, and there is no evidence that he completely identified with their goals. This approach is similar to his relationship to Zionism in general, namely, one of critical distance.[253] As we have seen, he perceived Palestine mainly as a mere instrument for advancing goals that were of more importance to him: the curing of the "sick Jewish soul" and the improvement of the status of the Jews in the eyes of non-Jews.

We can therefore conclude that Einstein's visit can be viewed as that of a Zionist tourist, or, more precisely, as that of a tourist *in* Zionism. His hosts presented him with a Zionist narrative, provided him with a highly organized and carefully coordinated tour of the country, showed him only the places they wanted him to see, did not inform him of internal Jewish conflicts, and kept him away from controversial individuals and locations. A large part of the responsibility for this lay with Einstein himself, as there is no evidence of his requesting to meet with more radical representatives of the Arab population or to tour areas that may have been less welcoming. But did he see the country through Zionist eyes? Due to his exceptionally independent mind, we can surmise that he saw the country through his own eyes. However, in light of the propaganda tour he was given, one could argue that while he was there, he wore heavily Zionist-colored glasses.

What was the personal impact of the trip on Einstein? Similar to other travelers, Einstein found travel disorienting.[254] There are numerous indications of this. He mixes up several place names in his diary entries. He confuses Singapore with Colombo, Haifa with Jaffa, and consistently misspells Hong Kong. Another indication of his disorientation is that he first writes "32nd December" instead of "1st January."

How did the long journey affect his relationship with his wife Elsa? The diary provides us with some revealing vignettes about their relationship. Literally, the first thing that actually happens on the trip is that he "[l]ost wife at border."[255] In Singapore, Elsa is "horrified" by the "lavatory with chamber pots and a big washtub."[256] Prior to their departure from Tokyo, the hasty packing leads to a quarrel.[257] From Kyoto, Einstein decides to travel to Osaka without Elsa. On his return, he has to face "great indignation by left-at-home wife."[258] In Penang, when the return to the ship is precarious due to rolling waves, Einstein remarks wryly, "Else[259] very frightened but had sufficient energy left to grumble." An indication that the long voyage

5. Einstein and Elsa aboard the S.S. *Kitano Maru* en route to Japan, October 1922 (courtesy of NYK Maritime Museum).

HISTORICAL INTRODUCTION

together had some impact on the degree of harmony between them can be seen in Einstein noting that he feels a "[s]ense of liberation" when he takes a walk by himself in Port Said while Elsa is ill.[260]

How did Einstein, who was an introvert and was not fond of grand festive occasions, cope with the onslaught of social engagements during his tours of the various countries he visited? The very day after his arrival in Japan, he was clearly overwhelmed by the hectic reception afforded him in Tokyo: "Arrival at hotel, completely exhausted among gigantic floral wreaths and bouquets. Still to come: visit by Berliners and burial alive."[261] At the end of a particularly eventful day in the capital, he arrives back at the hotel "dead tired."[262] The most colorful remark of this nature was undoubtedly recorded near the end of his Japanese tour: "But I was dead, and my corpse rode back to Moji."[263] By the time he had reached Palestine, Einstein's emotional distance from the social events was very apparent in his description of them as "comedies."[264] In Spain, he went a step further and described the social engagements as "punishment as always."[265]

No wonder, then, that during the outbound and return voyages, Einstein thoroughly enjoyed the isolation on board the ship and repeatedly noted in his diary how much he cherished the peace and quiet of the wide open sea. At the beginning of his return trip to Europe, he wrote to the acting chairman of the Nobel Committee for Physics, Svante Arrhenius: "And how conducive to thinking and working the long sea voyage is—a paradisiacal state without correspondence, visits, meetings, and other inventions of the devil!"[266] One could even go as far as stating that the ocean voyages were the primary motivation for Einstein's overseas travel and that the lecture tours were a secondary bonus and the price he had to pay for the moratoria of long sea travel. But the intensity of the Japan tour left Einstein with conflicting emotions as to whether he would embark on such an endeavor again in the future. By mid-December, he wrote to his sons from Kyoto: "I am determined not to gallivant around the world so much anymore; but am I going to be able to pull that off, too?"[267]

### Einstein's Scientific Research during His Trip

In contrast to all the public attention and hectic schedules of Einstein's visits along his way to and from the Far East, the quiet leisure time on board the ocean liners afforded him long stretches of uninterrupted time for research. The evidence for his work on the "gravitation-electricity problem" on the outbound journey is meager—he only twice mentions his musings and calculations. However, on his return voyage, he worked intensely on the technical intricacies of British astronomer Arthur S. Eddington's recent approach toward a unified field theory. Shortly after 10 January 1923, he completes the paper he has been working on.[268] Entries in his diaries and comments in his travel correspondence testify to Einstein's satisfaction about this opportunity to do some quiet thinking. For example, around the time he was finishing his manuscript, he wrote to Danish physicist Niels Bohr: "The voyage is splendid. I am enchanted with Japan and the Japanese and am sure that you would be, too. Moreover, such a sea voyage is a splendid existence for a ponderer anyway—like a monastery."[269]

### The Nobel Prize for Physics

On his arrival in Shanghai, Einstein learned from a telegram sent by Christopher Aurivillius, secretary of the Royal Swedish Academy of Sciences, that he had been awarded the Nobel Prize for Physics for the year 1921.[270] He had received the first hint that he was to become a recipient of the Nobel Prize from Svante Arrhenius, around 17 September.[271] Max von Laue had even advised Einstein to reconsider his trip to Japan: "According to reliable news I received yesterday, events could be taking place in November that would make your presence in Europe in December desirable."[272] While neither

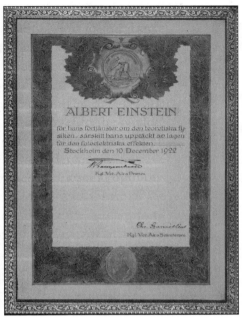

6. Nobel Prize in Physics, certificate, 10 December 1922 (with permission of the Albert Einstein Archives, the Hebrew University of Jerusalem).

of these communications mentioned the Nobel Prize explicitly, it is safe to assume that Einstein understood what was afoot. Nevertheless, he wrote Arrhenius that he was contractually bound to embark on the lecture tour and would not postpone the trip.[273] Intriguingly, Einstein made no note of the Nobel Prize award in his diary.

### The Reception of Relativity and the Impact of Einstein's Visits

Following the verification of his general theory of relativity, Einstein became an international celebrity overnight in November 1919. As we have seen, his travel in the ensuing years took him to numerous European locations. Einstein was motivated by two main factors: the dissemination of his theories and the reestablishment of international cooperation between the German and foreign scientific communities, which had been severed by World War I.[274] According to the German historian of science Jürgen Renn, "science became a messenger of international cooperation and Einstein its leading protagonist."[275] He embarked on these trips with such intensity that, by late 1920, he was ironically calling himself a "traveler in relativity."[276] Half a year later, he left the continent of Europe for the first time and visited both the United States and the United Kingdom.[277]

In this section, we will examine the reception of relativity in the countries Einstein toured on his way to the Far East and on the return leg of his journey. As some excellent historians of science have already done extensive research on relativity's reception in these locations, I summarize their insights here.[278] We will also explore what other forms of scientific and nonscientific impact Einstein's visits had on the countries he visited.

The theory of relativity made its initial appearance in China in 1917. It had been imported by Chinese students who had studied physics in Japan. Bertrand Russell presented a series on relativity in Beijing in 1921. Both his lectures and Einstein's brief visits led to "sharp increases in Chinese publications on relativity." Relativity was accepted in China "quickly and virtually without controversy." Historian of science Danian Hu attributes this "virtually unique" reception to its revolutionary image, which was attractive to Chinese intellectuals, and to "the absence of a tradition of classical physics in China."[279]

How were Einstein and relativity received in the most important destination of his voyage, Japan? Einstein's tour of the country was clearly an unbridled success. He was lionized wherever he went, his visit became a media sensation, and all of his scientific and popular lectures were extremely well attended and met with great enthusiasm. Interest in relativity was not restricted to the general public; the imperial court, political circles, and the business world were all fascinated by the illustrious visitor.[280] One factor that guaranteed the tour's positive outcome was the effective public relations campaign of the Kaizo-Sha publishing house. They instituted "a competitive cooperation with central and local newspapers," the special Einstein issue of *Kaizo* sold out, and the four-volume set of the Japanese translations of Einstein's collected writings (the first such publication worldwide) sold 4,000 copies during and after his tour.[281] Even though Einstein's tour of Japan was the only visit that can be defined as "a planned media event," the historian of science Thomas F. Glick has argued that the reception of his theories there was not substantially different than in other locations.[282] However, there was also somewhat of "a backlash against the publicity onslaught launched by Kaizosha," with some more conservative intellectuals being less enthusiastic about Einstein's visit.[283] There were also misconceptions about his theories. The term relativity "misled and amused many Japanese." Its Chinese characters were similar to the term for "sex between lovers," and this even led to discussions on the topic in the Imperial Diet.[284]

What was the impact of Einstein's six-week tour? One important consequence was a heightened interest in science in Japanese society.[285] As we have seen from Einstein's early contacts with Kuwaki and Ishiwara, relativity was introduced to Japan fairly early. His visit had a significant impact on the younger generation of Japanese scientists. However, there was also criticism that he had been monopolized by the older generation and that the younger members of the scientific community had not been given sufficient time for discussions with Einstein.[286]

The tour also had important political aspects. Einstein met with two prominent Christian reformers and labor activists, Toyohiko Kagawa in Kobe and Senji Yamamoto in Kyoto.[287] He was also in contact with more radical political groups. The letter the Japanese Proletarian Alliance sent Einstein was apparently censored by the authorities. The Alliance asked about his views "on the ----- imperialist government of Japan." The censored word was probably "aggressive." They also asked about his hopes for Japanese youth.[288] In his reply, he advocated the organization of Japanese labor. However, he warned against creating an opposition movement "merely for the sake of opposition," a stance expressed elsewhere by Einstein.[289] He also issued cautionary words about the dangers of militarism.[290]

The visit was also beneficial for German foreign policy. German science had already been revered in Japanese society prior to Einstein's visit.[291] As he was perceived as a German scientist throughout his tour, its success was also seen as a significant German achievement. The German ambassador Wilhelm Solf reported back to Berlin about "the extraordinarily beneficial effect of Einstein's visit for the German cause." However, he was concerned that the incident surrounding the statements of Maximilian Harden back in Berlin would harm the German cause.[292]

How was relativity received in Palestine? As in many other countries, the myth of the alleged incomprehensibility of the theory of relativity was an integral part of the reception of Einstein's theories there. At the time of his visit, there were only rudimentary beginnings of an academic infrastructure in the country. Even though Einstein lectured on relativity in both Jerusalem and Tel Aviv, the overall reception in the press and by political leaders was one of incomprehensibility. It seems that the very fact that the lectures were held was more important than their content, indicating that their national significance was more significant than their scientific importance. Some articles on relativity were published in the press during Einstein's visit, thereby contributing to some extent to the popularization of his theories in the *Yishuv*.[293]

What was the impact of Einstein's tour of Palestine? It is patently clear that the local Zionists presented the mainstream Zionist narrative about Jewish colonization efforts, the needs of the *Yishuv*, and relations with the local Arab inhabitants to their illustrious guest. The clear purpose of the itinerary of the visit was to present the achievements of the Jewish community in the best possible light. In alignment with more general developments in Zionist tourism of the time, the goal was "to entice Jews who were already interested in Zionism to visit and to ensure they carried away a favorable impression of the Zionist enterprise.[294] As we have seen, internal conflicts were downplayed. Centers of Arab nationalism, such as Jaffa, where violence had occurred less than two years previously, were also clearly avoided. Apart from some discussions with representatives of the British Mandate authorities, Einstein had minimal exposure to alternative narratives to the Zionist one. The Zionist establishment in Palestine emphasized the national, and even redemptive, dimension of Einstein's visit. In their speeches, their guest was glorified as "a genius" and even ascribed a quasi-messiah-like role for whom they had been waiting "for 2,000 years."[295] Intriguingly, in stark contrast to the national fervor that surrounded the focal point of his tour—his speech at the future site of the Hebrew University—his comments on the event in his diary are quite laconic and do not reveal his emotions about this momentous occasion. In his later accounts of his visit, Einstein does not mention the event at all.

It is evident that the Zionists used the tour for propaganda purposes. This was particularly blatant in the public speeches of Ben-Zion Mossinson, who made outrageous claims that Einstein was learning Hebrew and intended to settle in Jerusalem. It is possible to see these instances as a cynical exploitation of Einstein's ignorance of the Hebrew language.

The Zionist Organisation in London also stressed the utmost national importance of Einstein's visit. In welcoming him to Palestine, Chaim Weizmann bestowed on Einstein a national role he had not previously ascribed to him, that of national regenerator, both in Palestine and in the Diaspora.[296]

The Jewish community's reception and perception of Einstein can best be gleaned from the coverage of his visit in the local press. The superlatives lavished on Einstein were full of hyperbole, and the adulation was boundless. Einstein fever in the *Yishuv* reached a frenzy at the central event of his visit—the inaugural scientific lecture of the Hebrew University on Mt. Scopus. The daily *Ha'aretz* designated the celebrations "a national festival and a scientific festival." Einstein's mounting the stage to deliver his lecture was interpreted by the press as no less than a sacred moment. *Torah* was once more being dispensed from Zion. The distinguished visitor was perceived as the high priest in the new "temple of science that will be built in Zion."[297] Einstein had become a national icon. And it seems that he was aware of this, as can be discerned from his sketch of a halo around his head in his postcard to Zionist leader Arthur Ruppin.[298]

Relativity was first introduced to the Spanish scientific community in 1908 by the mathematician Esteban Terradas é Illa and the physicist Blas Cabrera. They were also mainly responsible for its dissemination after 1919. In the absence of a large enough number of research scientists in Spain, the reception of relativity was chiefly championed by mathematicians. Its main consumers were a newly emerging "scientific middle class," composed mainly of engineers and physicians. Many of the leading supporters were conservative Catholics. Popularizations for the general public rendered relativity a "culture-wide phenomenon engaging different levels of society and domains of discourse simultaneously." The main opponent in the scientific community to relativity was the astronomer Comas Solá.[299]

The impact of Einstein's visit to Spain was multilayered. The visit led to important changes in the popular image of science, which may have already been under way. As Einstein was perceived as being "a symbol of science's prestige," the intellectual elite became more inclined to welcome "pure science as part of culture." The scientific middle class aimed to displace the old political and cultural elites, who were embarrassed by the alleged "incomprehensibility" of relativity.[300] The visit also had a political dimension. In particular, Catalan nationalists tried to use the illustrious guest for their ideological causes.[301]

What general conclusions have historians of science drawn from the reception of relativity during this period?

As Renn has pointed out, Einstein's trips "enhanced the process of emancipation already under way in local scientific communities that were seeking a greater role for basic science in their societies."[302] Glick has concluded that the reception of relativity "obliged scientists to confront both themselves and their disciplines."[303] Consequently, the rise to fame of Einstein and his theories "triggered reflections about the status of science and of particular disciplines in a national community, as well as about the relation between local and global traditions of science."[304]

A newer approach to the question of the reception of scientific theories has looked at the nature of the transmission of scientific knowledge. Patiniotis and Gavroglu advocate an approach that moves away from traditional reception studies. Rather than focusing on the transmission of knowledge "from the 'center' *to* the 'periphery',", which is the perspective of reception studies, they prefer to focus on the "appropriation of ideas and practices *by* the 'periphery'."[305] Such an approach could benefit from the previously mentioned concept of "contact zones" by Mary Louise Pratt, in which "transculturation" takes place. In the process of transculturation, "the subordinate or marginal groups select and invent from materials transmitted to them by a dominant or metropolitan culture."[306] Detailed studies that apply the concept of the "appropriation" of relativity to the scientific communities in the countries Einstein visited on his tour of the Far East have yet to be written.

## Conclusions

What conclusions about Einstein's personality, his more general views and opinions, and the manner in which he traveled can we draw from his six-month-long trip halfway around the world?

We have seen that the ocean voyage almost instantly put Einstein in a frame of mind to be more introspective about his own identity and the identities of the various peoples he was about to encounter. The journey also forced him to confront his own multiple identities: ethnically as a Jew, nationally as a German and a Swiss, continentally as a European, and hemispherically as a Westerner.

The diary has provided an insight into how Einstein traveled when he was abroad. In sharp contrast to his intra-European trips, he exhibited far greater passivity when journeying overseas. He was quite happy for his respective local hosts to take care of almost all the arrangements. He was prepared to surrender his autonomy—to a large extent—in exchange for greater convenience. The extended trips also granted him new opportunities to be absent from "nerve-wracking" Berlin for longer periods.

The manner in which he traveled created a fascinating juxtaposition. Even though he and Elsa were the only participants in the journey and, therefore, the trips should theoretically have taken on the character of those suited more for individual travelers, the nature of the visits were actually similar to package deals for tourists.

Another intriguing question is: who was served more by Einstein's visits, his hosts or he himself? In Japan, he was definitely beholden to the tight and exhausting schedule of the Kaizo-Sha publishing house. However, at the same time, he also used the tour to propagate and disseminate his scientific theories. The long absence from a precarious political situation in Berlin was also a great bonus for Einstein. In Palestine, the balance leaned perhaps more in his hosts' favor. For the Zionists, the visit of the most prominent non-Zionist Jewish personality was an amazing propaganda coup. For Einstein, the tour was mainly a means to satisfy his curiosity about the country whose development he had been supporting for some years. It also afforded him the opportunity to be present at one of the defining moments of the Hebrew University, the Jewish cause that meant the most to him.

As for conclusions about Einstein's beliefs, we can determine the following. We have seen in numerous instances that Einstein espoused geographical determinism. With the temperatures constantly rising, he states that he is "convinced that the Greeks and the Jews of classical antiquity lived in a less sluggish atmosphere. It is no coincidence that the zone of vibrant intellectual life has since slid northward. So much more convenient for vegetating."[307]

The diary also reveals his preference for genetics over environment. While visiting Hong Kong and encountering its Jewish community, he is impressed with the similarity between Jews of Middle Eastern and European heritage. One can place this favorable disposition toward genetics in the context of Einstein's biological worldview.[308] Ironically, his belief in nature over nurture actually contradicts his advocacy of geographic determinism.

Another intriguing aspect of Einstein's beliefs revealed in the diary is his attitude toward respectability and morals. In numerous instances, he expresses his admiration for stench and dirt (mainly in China), yet he also holds in high esteem the cleanliness and impeccable manners of the Japanese. This contradiction is riveting in light of Einstein's overall disdain for bourgeois respectability.

The diary has revealed some fascinating contradictions in Einstein's personality. Even though he had a strong disdain for nationalism, he firmly believed in national character. In spite of his great sympathy for the oppressed and even, at times, feeling ashamed of being complicit in their exploitation, he clearly held Orientalist views, advocated an "enlightened" form of colonialism, and did not always accept the basic humanity of the local populaces. Even though he felt discomfort with the plight of some of the indigenous peoples he encountered, he subscribed to some disturbing racist beliefs, particularly with regard to the alleged biologically caused intellectual inferiority of the Japanese and Chinese, and he even feared Chinese racial domination.

The journal also helps us understand in part how Einstein perceived of himself. He often presents himself as having to confront multiple attacks from a manifold of challengers. In this, he symbolically adopts the role of a protagonist, one who has set out on a quest—perhaps even a crusade or mission—to disseminate relativity (and by implication, Western civilization) to the "natives," even in the face of their many attempts to "assail" him. As his Japanese host put it, his arrival in Tokyo was "like welcoming a general returning from a victorious campaign."[309] Einstein's self-presentation as a hero aligns nicely with one of his favorite quotes from a German poem he used to describe himself self-mockingly: "But the valiant Swabian is not afraid."[310]

The many instances of national stereotyping and racist remarks expressed in the diary beg the question of how these beliefs align with the common public perception of Einstein the humanitarian. Indeed, how can they possibly be compatible with his more enlightened public statements and support for progressive causes? How could this "gentle icon of humanist values"[311] possibly be so prejudiced?

One obvious explanation is that this travel diary was never intended for publication. It was either written by Einstein as an aide-mémoire for future reference, or, as mentioned previously, for the sake of his stepdaughters back in Berlin. This would account for the discrepancy between his more tolerant and guarded public statements and his candid and (at times) offensive private notes. The travel diary allows Einstein to explore his more irrational and instinctual side and to be freer in his expression of his personal prejudices.

Another explanation is Einstein's pronounced elitism with regard to intellect. His humanism ends when he perceives a nation as intellectually inferior.[312] Furthermore, it could be argued that in spite of his public advocacy of human rights, it was science, not humanity, that lay at the center of Einstein's universe.

A further explanation is that the public image of Einstein as a notable humanitarian (as opposed to his political support for progressive and left-wing causes) developed at a later stage than the writing of this journal. It became particularly prominent in the post-World War II era.

On his trip to the Far East, one of the main challenges for Einstein was to recognize himself in the Other and his own Otherness in himself. We have seen that he did not always manage to do so. Yet we should not be too judgmental in our assessment. It is also unsettling for us to face the more troubling sides of this cultural icon, and thereby acknowledge the more disquieting aspects of our own personalities. And in today's increasingly less-tolerant world, it is still a particularly difficult challenge to recognize ourselves in the Other and their Otherness in ourselves.

# Travel Diary
# Japan, Palestine, Spain[1]

6 OCTOBER 1922–12 MARCH 1923

6. Oktober. Nachtfahrt im überfüllten Zuge
nach Wiedersehen mit Besso und Chavan.
An Grenze Frau verloren. Sonnenaufgang
kurz vor Ankunft in Marseille. Silhouetten
erster flacher Häuser, von Pinien umgeben.
Marseille enge Gassen. Üppige Frauen. Vegetier-
Leben. Wir wurden von bieder scheinendem Jüngling
ins Schlepptau genommen, in grässlichem Wirtshaus
nächst der Bahn abgesetzt. Käfer im Morgenkaffee.
Weg ins Schiffsbüro und zum alten Hafen nebst
altem Stadtquartier. Am Schiff energische Ab-
fertigung des Spitzbuben, der beleidigt abfuhr, nach
federloser Fahrt zum Hafen auf Kofferwagen über
ungeheuer holpriges Pflaster Marseilles. Dort
nur mündliche Gepäckrevision. Freundlich
empfangen von Schiffsoffizier. In Kabine behaglich eingerichtet. Jungen japanischen Arzt kennen
gelernt, den ein Münchener Mediziner mit seinem
flammenden Ultimatum an die Gelehrten
der Entente-Länder hinausgeworfen hatte.
8 X. Beschränkter Sonntag im Hafen. Mittags
Abfahrt bei heller Sonne. Fast nur Engländer
und Japaner auf dem Schiff. Stille, feine Gesellschaft.
Nach Ausfahrt aus dem Hafen wunderbarer Blick

6th October. Night trip in overfilled train after reunion with Besso[2] and Chavan.[3] Lost wife[4] at border. 7th Oct. Sunrise shortly before arrival in Marseille. Silhouettes of austere flat houses surrounded by pines. Marseille, narrow alleyways. Voluptuous women. Vegetative living. We were taken in tow by seemingly honest youth, dropped off at ghastly inn by the railway station. Bugs in morning coffee. Made our way to the shipping company and to the old harbor near the old city quarter. At the ship,[5] energetic dispatch of the scamp, who departed in a huff after a jarring ride to the harbor on a luggage cart over dreadfully bumpy pavement of Marseille. There, just a cursory luggage review. Friendly welcome by ship's officer. Comfortably settled in cabin. Met young Japanese physician whom a Munich medical doctor had thrown out with an inflammatory ultimatum to scholars of the Entente countries.[6]

8th Oct. Sedate morning in the harbor. Joyously greeted by rotund Russian Jewess, who recognizes me as a fellow Jew. Noon, departure in bright sunshine. Virtually only Englishmen and Japanese on board the ship. Quiet, decent company. After leaving the harbor, wonderful view

7. The S.S. *Kitano Maru* (courtesy of NYK Maritime Museum).

auf Marseille und die es rahmenden Hügel
Dann an größeren schroffen Kalkfelsen vorbei. Küste wedelt
langsam links zurück. Gespräch mit europäisiertem
japanischem Arzt Miyake aus Fukuoka.

   Nachmittags 4 Uhr Rettungsprobe. Alle Passagiere
müssen – mit denen in der Kabine aufbewahrten Rettungs-
gürteln angethan – an der Stelle zur Musterung
erscheinen, an der das für sie bestimmte Rettungsboot
im Falle der Gefahr zu benennen ist. Schiffsmannschaft
(lauter Japaner) freundlich, genau ohne Pedanterie, ohne
individuelle Note. Er (der Japaner) ist unproblematisch, impersönlich,
füllt heiter die ihm zugefallene soziale Funktion
aus, ohne Pretension, aber stolz auf seine Gemein-
schaft und Nation. Das Aufgeben seiner traditionellen
Eigenart zugunsten der europäischen gebot nicht an
seinem Nationalstolz. Er ist impersönlich, aber nicht
eigentlich verschlossen; denn als vorwiegend soziales
Wesen scheint er für seine Person nichts zu besitzen,
das zu verschließen oder zu verbergen er das Bedürfnis
haben könnte.

9. X. 4 Uhr Morgens großer Krach. Ursache Schiffsreinigung.
Große Reinlichkeit an Menschen und Sachen. Das Schiff
ist wie abgeschleckt. Es wird schon bedeutend
wärmer. Die Sonne erquickt mich und nimmt

of Marseille and its surrounding hills. Then pass by harsh craggy chalk cliffs. Coast slowly recedes away to the left. Conversation with Europeanized Japanese physician Miyake from Fukuoka.[7]

Afternoon, 4 p.m., safety drill. All passengers—wearing the life-preservers stored in their cabins—must report for inspection at the location of the lifeboat designated for them in the event of emergency. Crew (all Japanese) friendly, precise without being pedantic, don't stand out as individuals. He (the Japanese) is unproblematic, impersonal, he cheerfully fulfills the social function which befalls him without pretension, but proud of his community and nation. Forsaking his traditional ways in favor of European ones does not undermine his national pride. He is impersonal but not actually withdrawn; for, as a predominantly social being, he seems not to possess anything individually that he would have the need to be taciturn or secretive about.

9th Oct. 4 a.m. major racket. Cause: scrub-down of vessel. Great cleanliness of people and things. The ship appears as if licked clean. It is already becoming significantly warmer. The sun revitalizes me and removes

die Kluft zwischen „ich" und „es" weg. Ich beginne mit der Lektüre von Kretschmers Körperbau und Charakter. Wunderbare Schilderung der Temperamente und ihres physischen Habitus. Kann viele Mitmenschen so objektivieren, mich selber aber nicht, weil hoffnungsloser Mischtypus. Gestern las ich Bergsons Buch über Relativität und Zeit. Merkwürdig, dass ihm nur die Zeit aber nicht auch der Raum problematisch ist. Er scheint mir mehr sprachliches Geschick als psychologische Tiefe zu haben. Bei der Objektivierung des Psychisch-Gegebenen macht er sich wenig Skrupel. Er scheint aber die Relativitäts-Theorie sachlich zu begreifen und setzt sich mit ihr nicht in Gegensatz. Die Philosophen tanzen beständig um den Gegensatz Psychisch-real und Physikalisch-Real herum und unterscheiden sich nur durch Wertungen in dieser Beziehung. Entweder erscheint ersteres als „bloßes Individualerlebnis" oder letzteres als „bloße Gedankenkonstruktion". Bergson gehört zur letzteren Gattung, objektiviert aber unvermerkt in seiner Weise.

Habe wieder über das Problem Gravitation-Elektri-

the gulf between "ego" and "id." I begin reading Kretschmer's *Physique and Character*.[8] Wonderful description of temperaments and their physical habitus. I can thus categorize many of my fellow beings but not myself, because I'm a hopeless hybrid. Yesterday I perused Bergson's book on relativity and time.[9] Strange that time alone is problematic to him but not space. He strikes me as having more linguistic skill than psychological depth. He is not very scrupulous about the objective treatment of emotional factors. But he does seem to grasp the substance of relativity theory and does not oppose it. Philosophers constantly dance around the dichotomy between the psychologically real and the physically real, and differ only in value judgments in this regard. Either the former appears as a "mere individual experience" or the latter as "a mere construct of thought." Bergson belongs to the latter category but objectifies in *his* way without noticing.

I've been thinking about the gravitation-electricity

zität nachgedacht. Finde, dass Weyl darin
Recht hat, dass ein g-Feld bezw. eine Invariante
als losgelöst vom Elektrischen keine Realität
hat, also auch nicht mathematisch
nicht objektiviert werden kann. Ich
glaube aber, dass die endgültige
Lösung weiter von Riemann abliegt
wie bei Weyl, glaube auch, dass abgelöst
von Elektromagnetie dem Elementargesetz
der vektoriellen Parallelverschiebung unmittel
bar nichts entspricht, und dass deren Form
lismus als Grundlage der Theorie (über Riemann hinaus) keine
objektive Berechtigung hat. Dagegen scheint
es mir doch möglich, dass man an der
Feldtheorie wird festhalten können, ob
auch am Ausdruck der Naturgesetze durch
Differentialgleichungen, scheint zweifelhaft.
16. X. Heute Morgen links als Stromboli
vorbeigefahren. Dampfwolken. Prachtvoll in
Morgensonne.

nur einige Vulkan-Kegelragen aus dem Meer

problem again. I find that Weyl is right that a $g_{\mu\nu}$ field, or an invariant *ds* independent of electricity, has no reality, therefore cannot be mathematically objectified either. But I do think that the ultimate solution is further away from Riemann than Weyl's approach and also think that ⟨independent of the electromagnetic⟩ nothing directly corresponds to the elementary law of the parallel displacement of a vector and that this formalism has no objective justification as the basis of the theory beyond Riemann. However, it does seem possible to me that one will be able to hold on to field theory; whether the expression of natural laws by differential equations will also be retainable appears doubtful.[10]

10th Oct. This morning sailed past Stromboli on the left-hand side.[11] Clouds of steam. Magnificent in the morning sunlight.

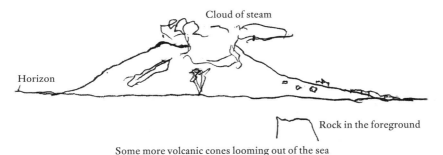

Some more volcanic cones looming out of the sea

## 3

Laue berauschende Luft. Stahlfarbenes Meer.
Italienische Festlandküste diffus wolkig.
Japanerinnen krabbeln mit Kindern herum
Sehen blumig und verwundert aus, fast
wie ~~schematisch~~ stilisiert. Schwarzäugig, schwarzhaarig
grossköpfig trippelnd.

War gestern noch erschüttert von Lektüre von
Kretschmer. Fühlte mich mit Zange gepackt. In Gleich-
gültigkeit verwandelte Hypersensibilität. In Jugend
innerlich gehemmt und weltfremd. Glaswände
zwischen Subjekt und andern Menschen. Unmotiviertes
Misstrauen. Papierene Ersatzwelt. Asketische Ver-
wandlungen. Ich bin Oppenheim für das Buch
dankbar. —

Herrliche Fahrt am Mittag durch die Meerenge
von Messina. Kahle ernste Berglandschaft zu
beiden Seiten. Städte ebenfalls ernst, Vorherrschen
der Horizontale. Niedere flache weisse Häuser.
Orientalischer Gesamteindruck. Temperatur steigt
unablässig. Bin überzeugt, dass die Griechen
und Inder des klassischen Altertums in weniger
erschlaffender Atmosphäre lebten. Es ist kein
Zufall, dass die Zone lebhaften geistigen Lebens
seither nordwärts gerutscht ist. Unser angenehmer

Balmy, intoxicating air. Steel-colored sea. Hint of Italian mainland through the fog. Japanese women crawling about [on deck] with children. They look ornate and bewildered, almost as if ⟨schematic⟩ stylized. Black-eyed, black-haired, large-headed, scurrying.

Yesterday, was still unsettled by reading Kretschmer. Felt as if grabbed by a vice. Hypersensitivity transformed into indifference. During adolescence, inwardly inhibited and unworldly. Glass pane between subject and other people. Unmotivated mistrust. Substitute paper world. Ascetic impulses. I am so grateful to Oppenheim for the book.[12]

Splendid passage through the Strait of Messina at midday. Stark, austere mountainous landscape on both sides. Towns likewise austere, predominance of the horizontal. Low, flat white houses. Overall impression: Oriental. Temperature rising relentlessly. I'm convinced that the Greeks and the Jews of classical antiquity lived in a less sluggish atmosphere. It is no coincidence that the zone of vibrant intellectual life has since slid northward. So much more convenient

3v

das Vegetieren. Das Streben nach Zufrieden-
heit hier leichter zu befriedigen, weil es selbst
zum lebhaften Wünschen fast schon zu arm ist.
11.X. Sonniger Tag. Himmel weißlich. Meer etwas bewegt.
Glaube jetzt, dass Seekrankheit auf Orientierungs-
schwindel beruht, nicht auf Änderungen der schein-
baren Schwere nach Richtung und Größe.
12.X. Strahlender Tag. See ruhig, fast windstill,
Atmosphäre ganz durchsichtig. Horizont scharf. Fast
windstill. 7½ Uhr Felsgebirge von Kreta, steil ins
Meer abfallend. Abends wunderbarer Sonnenuntergang — purpurn mit fein beleuchteten
schmalen Windwolken. Dann tausender
Sternenhimmel mit stark heraustretender
Milchstr. bei lauer Luft. Mittags Ankunft
Port Said. Vormittags Gespräch mit Italie-
und japanischen Touristen. Beide stark
europäisiert — bis zum Ausdruck
des Gesichtes. Ersterer sehr vorsichtig und
realistisch. Abends Gespräch mit Hamburger
Exporteur. Zaghaft und nüchtern.
13. Mittags Port Said. Wasser grün, bevor Küste
sichtbar. Im Mittelmeer tief blau. Lange
künstliche Dämme mit Drahtblöcken ...

for vegetating. The quest for contentment is more easily satisfied here, because it is already almost too hot, even for active wishing.

11th Oct. Sunny day. Whitish sky. Sea somewhat restless. I now think that seasickness is based on dizziness caused by lack of orientation, not directly on the apparent changes in gravity, according to direction and magnitude.

12th Oct. Radiant day. Sea quiet, almost windless. Atmosphere completely transparent. Distinct horizon. Almost windless. At half past seven in the morning, rocky mountains of Crete visible, sloping steeply into the sea. In the evening, wonderful sunset—purple with finely illuminated narrow wind-swept clouds. Then, bright starry sky with prominent Milky Way in the balmy air. ⟨Noon arrival Port Said.⟩ In the morning, conversation with Ishii[13] and Japanese lawyer. Both strongly Europeanized—down to the facial expressions. The former is very cautious and realistic. In the evening, conversation with exporter from Hamburg betw[een] 60 & 70, shrewd and level-headed.

13th. Midday. 3 p.m., Port Said. Water [turned] green before the coast became visible. In the Mediterranean, deep blue. Long artificial dams with irregularly shaped blocks of stone. Houses,

soweit vom Meere sichtbar europäisch. Malerisch
ägyptisches Segelschiff vom Typus

Im Hafen Gewimmel von Kähnen mit
schreienden und gestikulierenden Levantinern
in allen Schattierungen, die sich auf unser
Schiff stürzen. Wie von der Hölle ausgespien
Ohrenbetäubender Lärm. Das obere Deck
in Bazar verwandelt, aber niemand
kauft etwas. Nur einige hübsche athletische
junge Wassersager haben Erfolg. Sonnenuntergang
Himmel lokal und sehr intensiv geritzt, wie
flammend. Auf der Gegenseite Mauern u. Gebäude
in jener scharfen Farbe, die oft auf Tropenbildern
festgehalten. Abends Unterhaltung mit franz. Beamten
aus Siam. Begegnung mit japanischen Schwesterdampfer. Begeisterung
14. Erwachen bei Kanalfahrt durch Wüste.
Kühle Temperatur. Palmen, Kamele. Stechend
gelbe Farbe. Blick oft auf ungeheure Sandfläche,
von Büschelchen Kraut unterbrochen. Dampfer
Begegnung im Kanal. Weite Wüstenblicke. Grüner
Bittersee. Hügelufer in glänzender Beleuchtung

as far as one could see from the sea, European in style. Picturesque
Egyptian sailboat of the type

In the harbor, a swarm of rowing boats with screaming and gesticulating Levantines of every shade, who lunge at our ship. As if spewed
from hell. Ear-deafening din. The upper deck transformed into a bazaar, but nobody purchases anything. Only a few handsome, athletic,
young fortune-tellers are successful. Bandit-like filthy Levantines,
handsome and graceful to look at. Sunset, sky reddened *locally* and
very intensely, as if blazing. On the facing side [of the harbor], walls
and buildings, one of those fierce colors that are often depicted in
paintings of the tropics. In the evening, conversation with Fren[ch]
civil servant from Siam. Encounter with Japanese sister ship. Nationalistic fervor of the crews. Japanese is in love with his country and
his people.

14th. Wake up during canal passage through desert.[14] Cool temperature. Palms, camels. Glaring yellow color. Frequent views of immense stretches of sand, intermittently broken up by tufts of weed.
Encounter steamer in the canal. Wide desert vistas. Green Bitter
Lake.[15] Hilly shore in radiant light.

Wunderbar klare Luft. Dann letzter Teil des
Suez-Kanals Bei Mündung Stadt Suez
Villenviertel der Kanal Verwaltung sehr hübsch
Veranda-Häuschen mit Palmen. Freundlicher
Anblick nach so viel Wüste. Beidseitig
Felsengebirge. Ausläufer des Sinai Arabische
Kleinenflotte segeln herbei. Sind schöne
Wüstensöhne, stämmig, strahlend schwarzaugig
und männlicher als in Port Said.

Bucht verbreitert nach Abfahrt. Sonne ver-
schwindet nach 6 über ägyptischen Wüsten-
Gebirge. Lokale rotviolette Färbung, bei prächtiger
Silhouette. Ferner rötlichgelber Widerschein
im Osten. Dann wundervolle sternhelle
Nacht. Milchstr. noch nie so schön gesehen.
Flecken mit deutlicher Begrenzung. Längliche
Flecken treten deutlich seitlich heraus
Nachts nackt mit Ventilator. Nicht beschwerlich
15. Im roten Meer bei bedecktem Himmel
ohne Sicht bis Ufer. Himmel leicht bedeckt.
Morgens kurzer Regen. Treibhaustemperatur.
Mittags an zwei flachen kleinen Koralleninseln
vorbei.
16. Zwei Haifische neben Dampfer

Wonderfully clear air. Then, last part of the Suez Canal. At the mouth, city of Suez, villa neighborhood of the canal administrators, very pretty. Small homes with verandas and palm trees. Welcome sight after so much desert. Barren rocky mountains on both sides. Extensions of the Sinai. Arab minor merchants sail up. They are handsome sons of the desert, brawny, shining black eyes and better mannered than in Port Said.

Gulf widens after departure. Sun vanishes after 6 over the Egyptian desert mountains. Local reddish violet hue with magnificent silhouette of the mountain range, which resembles the Ütliberg.[16] Additionally, reddish yellow reflection in the east. Then, wonderful starlit night. Have never seen the Milky Way so beautiful. Spots with distinct borders. Elongated spots can be seen distinctly clearly, emerging from the disk laterally. At night, naked with fan. Not troublesome

15th. In Red Sea under overcast skies without shore being visible. Lightly cloudy skies. Brief rain shower in the morning. Greenhouse temperature. At noon, passed by two small flat coral islands.

16th. Saw two sharks next to steamer

gesehen mit gewaltiger Rückenflosse und
Schwanzflossen, Temperatur zunehmend,
aber sehr erträglich. Wunderbarer Sonnenuntergang.
Widerschein im Osten mit rötlicher Mehrotyphbent-
Östlicher Himmel blaugrau, weiter oben rötlich
Venus glänzt wunderbar und spiegelt sich
im Meer.

17. Abends bei hellem Himmel mächtiges
Ferngewitter auf der arabischen Seite. Fast
jede Sekunde Blitz auf kleinem Winkelraum.
Darauf erhob sich starker Wind bei steigendem
Seegang.

18. Bei starkem Seegang (morgens) Ausgang des
roten Meeres erreicht. Arabische Felsengebirge
sichtbar; aus dem Meer ragende Einzelberge
glänzen rötlich in Morgensonne.

19 Darmkatharrh mit abscheulichen Hundeboden.
   Japanischer Professor eilt zu Hilfe

20. Wieder einigermassen flott. An Somalinküste
(Kap) vorbeigefahren. Herrlich in Nachmittags-
sonne beleuchtetes Gebirge. Frische Brise
mit mässigem Seegang.

25. Korallen-Inseln passiert mit Palmenhainen.
   In der Nacht Tropenregen. Jetzt 9½ Uhr fener

with huge dorsal and tail fins. Flying fish too. Temperature rising but very tolerable. Wonderful sunset. Reflection in the east with reddish surface of the sea. Eastern sky blue-gray, reddish farther up. Venus shines beautifully and is reflected in the sea.

17th. In the evening, mighty distant storm on the Arabian side under bright sky. Strokes of lightning onto small angular area almost every second. After that a strong wind rose up with swelling seas.

18th. Reached the exit of the Red Sea in strong swell by the morning. Arabian mountain range visible; lone-standing mountains looming out of the sea, glowing red in the morning sun.

19th. Enteritis with ghastly hemorrhoids.

Japanese professor comes to the rescue.[17]

20th. More or less shipshape again. Steamed past the Somali coast (cape).[18] Splendidly illuminated mountains in the afternoon sun. Brisk breeze with moderate swell.

25th. Passed by coral islands with palm groves. At night, tropical rain. Now 9:30, distant

Tropenregen sichtbar (bei wechselnd bedecktem Himmel).

Bei Annäherung an den Aequator nahm vom
arabischen Golf an die Bewölkung immer zu.
Im Oktober liegt die meist bestrahlte ~~Jahreszeit~~
Partie der Erdoberfläche am Aequator. Dort steigt
die mit Feuchtigkeit gesättigte Luft auf und
sondert dabei Wasser aus. Von beiden Seiten
(Nord und Süd) strömt Luft nach (Winde
der subtropischen Zone, abgelenkt durch Erddrehg.)
Jahreszeiten verschieben die Zone maximaler
Erwärmung und damit den ganzen Thermen –
Komplex nach Norden oder Süden. Ferner verschiebt
Land das Temperaturmaximum im Gegensatz
zu Meer, weil es viel stärker erwärmt wird.

Die starken aufsteigenden Luftströme begün-
stigen auch Gewitterbildung. Wir sahen oft
Wetterleuchten ohne oder bei schwacher
Wolkenbildung.

Im arabischen Golf viele Haifische
und fliegende Fische. Im offenen Meer,

tropical rain visible (with variably cloudy skies).

On approach to the equator, cloudiness increased from the Arabian Gulf onward. In October the most irradiated ⟨layer⟩ part of the Earth's surface lies at the equator. There moisture-saturated air rises, thereby eliminating water. Air flows in afterward from both (north and south) sides (winds of the subtropical zone, diverted by the Earth's rotation). Seasons shift the zone of maximum warming and hence the entire complex of phenomena northward or southward. Additionally, *land* intensifies the maximum temperature, as opposed to the sea, because it heats up to a greater degree.

The strong rising currents of air also encourage formation of storms. We often saw sheet lightning without any or with only weak cloud formation.

In the Arabian Gulf[,] many sharks and flying fish. Nothing of the kind was to be seen on the open sea,

das mehrere tausend Meter tief ist, sah man nichts dergleichen. Es dringt wenig Licht auf den Meeresboden, also schwacher Pflanzenwuchs am Boden, geringe Fauna unten, und erst recht oben.

In der Nacht Schiffs-Sirene. Glaubten Unglück. Waren aber nur Schallsignale wegen Undurchsichtigkeit der Luft bei starkem Regen *sehr dem* Falle von Schiffsbegegnungen. Temperatur sehr erträglich, nur in Kabine sehr heiss (zwischen sonnen-beschie-nener Schiffswand und einem an Maschinen Raum grenzenden Gang. Habe viel unter Unpässlichkeit zu leiden. Japanischer Arzt stets hilfsreich. Ich wurde auf dem Schiff viel mit und ohne Leute photographiert, hauptsächlich von Japanern.

Gestern habe ich die elektromagnetischen Vakuum-Gleichungen im Sinne der Weyl-schen Geometrie umgerechnet, in der Hoffnung, einen Ausdruck für die Stromdichte zu finden. Es kommt aber unbrauchbares Resultat $\varphi''^x \varphi_x$ heraus.

which is several thousand meters deep. Little light penetrates to the sea floor, so weak plant growth at the bottom, scant fauna below, even less so above.

During the night, ship's siren. Thought it was an accident. But was just acoustic signals owing to lack of visibility in heavy rainfall, in case of encounters with vessels. Temperature very tolerable, only inside the cabin very hot (between sunlit wall of ship and a corridor bordering on the engine room. Often feel unwell; Japanese doctor[19] always helpful. On the ship I am frequently photographed, with and without people, mainly by Japanese.

Yesterday I worked on the electromagnetic vacuum equations $\left( \frac{\partial \varphi^{\mu\nu} \sqrt{-g}}{\partial x_\nu} = 0 \right)$ according to Weyl's geometry in the hope of finding an expression for the current density. But a useless result $\varphi^{\mu\alpha} \varphi_\alpha$ emerges.

28. Gestern Abend näherten wir uns unter
erheblicher Verspätung ~~Hook~~ Colombo. Bevor
die Küste sichtbar war kamen wir in schweres
Tropengewitter mit Wolkenbruch, sodass
das Schiff halten musste. Als es sich um
9 Uhr aufhellte, zeigte es sich dass wir dem
Hafen nahe waren. Ein Lotse kam im Ruderschiff
heran und bald landeten wir neben einem
andern japanischen Dampfer. Wir sahen hier
zum ersten Mal einen älteren Inder, feines
somebues Gesicht mit grauem Bart, der uns
zwei Telegramme brachte und — um Trinkgeld
flehte. Noch andere Inder sahen wir, braune
bis schwarze schmige Gestalten mit aus-
drucksvollem Gesicht und Körper und
ergebenen Wesen. Sie sehen aus wie in
Bettler verwandelte Edelleute. Unaussprech-
lich viel stolzes und Deprimierendes ist da
vereinigt.

Heute Morgen um 7 Uhr setzten wir
aufs Land über und besahen uns mit der Plätze
zusammen das Hindu - Viertel von Colombo
und einen buddhistischen Tempel. Wir fuhren
in einzelnen Wägelchen, die von barfüßigen

28th. Yesterday evening we approached ⟨Honk⟩ Colombo with considerable delay. Before the coast came into view we got caught in a severe tropical storm with a cloudburst, forcing the ship to stop. When it brightened up around 9 o'clock, it turned out that we were near the harbor. A pilot drew near in a rowboat, and we soon docked next to another Japanese steamship. We saw here for the first time an elderly Indian, fine, distinguished face with gray beard, who brought us two telegrams and—begged for a tip. We saw other Indians as well, brown to black sinewy figures with expressive faces and bodies and humble demeanor. They look like nobles transformed into beggars. Much unspeakable pride and downtroddenness are united there.

This morning at 7 a.m. we went on land and, together with the Du Plâtres, viewed the Hindu quarter of Colombo and a Buddhist temple.[20] We drove in individual little carts that were drawn on the double by Herculean

und doch so feinen Menschen im Laufschritt
gezogen wurden. Ich habe mich sehr geschämt
an einer so abscheulichen Menschenbehand-
lung mitschuldig zu sein, konnte aber
nichts ändern. Denn diese Bettler in Königs-
gestalt stürzen doch scharenweise und
auf jeden Fremden, bis er vor ihnen kapi-
tuliert hat. Sie verstehen zu flehen und
zu betteln, dass einem das Herz wackelt.
Auf den Strassen des Eingeborenen-Viertels
sieht man diese feinen Menschen, wie sie
ihr primitives Leben verbringen. Sie machen
bei aller Feinheit den Eindruck, als wenn
das Klima sie verhinderte, mehr als eine
Viertelstunde nach rückwärts und vorwärts zu denken. In grossem Dreck und
respektablem Gestank hausen sie zu
ebener Erde, tun wenig und brauchen
wenig. Einfachster wirtschaftlicher Kreislauf
des Lebens. Viel zu zusammengepfercht,
um dem Individuum ein Sonderdasein
zu ermöglichen. Halbnackt zeigen sie die
feinen und doch kräftigen Leiber und
die feinen, geduldigen Gesichter. Nirgends

and yet so refined people. I was very much ashamed of myself for being complicit in such despicable treatment of human beings but couldn't change anything. Because these beggars in the form of kings descend in droves on any foreigner until he has surrendered to them. They know how to implore and to beg until one's heart is shaken up. On the streets of the indigenous quarter one can see how these fine people spend their primitive lives.[21] For all their fineness, they give the impression that the climate prevents them from thinking backward or forward by more than a quarter of an hour. They live in great filth and considerable stench down on the ground, do little, and need little. Simple economic cycle of life. Far too penned up to allow any distinct existence for the individual. Half-naked, they reveal their fine and yet powerful bodies and their fine, patient faces. Nowhere

Geschrei wie bei den Levantinern in Port Said.
Keine Brutalität, kein marktschreierisches
Leben, sondern stilles, ergebenes Vor-sich-Hinleben,
das einer gewissen Heiterkeit nicht entbehrt.
Wenn man diese Menschen recht ansieht, kann
man an Europäern kaum mehr Freude haben,
weil sie verweichlichter und brutaler sind und so viel
roher und Begehrender aussehen — und darauf
beruht leider ihre praktische Überlegenheit,
ihre Fähigkeit, grosse Dinge in Angriff zu
nehmen und durchzuführen. Ob wir in diesem
Klima nicht auch so würden wie die Tuller?

   Im Hafen lebhaftes Treiben. Hockerlische mit leuchtend schwarzen Leibern Arbeiter
besorgen die Ladung. Taucher produzieren sich in
ihren halsbrecherischen Künsten. Immer dieses
Lächeln und Sich-Hingeben für das schnöde
Geld und satte Menschen, die gemein genug
sind, um sich dessen freuen zu können. Um
12½ Uhr fuhren wir los, in die regnerische Wasser-
wüste hinaus. Ceylon ist ein Pflanzenparadies
und doch ein Schauplatz jammervollen Men-
schendaseins

31. Gestern war Geburtstag des Mikado. Feier auf
oberem Deck am Vormittag. Bansai und Absingen

shouting like the Levantines in Port Said. No brutality, no market crying existence, but quiet, acquiescent drifting along, albeit not lacking in a certain lightheartedness. Once you take a proper look at these people, you can hardly take pleasure in the Europeans anymore, because they are more effete and more brutal and look so much cruder and greedier—and therein unfortunately lies their practical superiority, their ability to take on grand things and carry them out. Wouldn't we too, in this climate, become like the Indians?

In the harbor, lively bustle. Herculean laborers with shiny black bodies take care of the cargo. Divers perform their neck-breaking skills. Always that smile and self-effacement for filthy money and self-satisfied people who are mean enough to gloat about it. At 12:30 we set out into the rainy desert of water. Ceylon is a paradise of vegetation and yet a scene of woeful human existence.

31st. Yesterday was the Mikado's birthday.[22] Celebration on the upper deck before noon. Banzai and singing

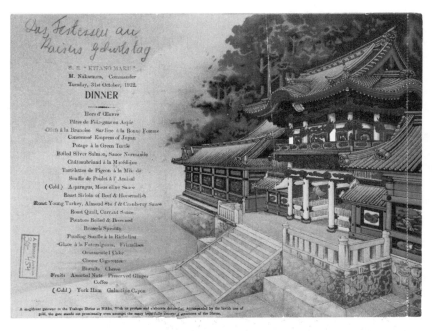

8. Dinner menu for Japanese Emperor's birthday, on board the S.S. *Kitano Maru*, 31 October 1922 (with permission of the Albert Einstein Archives, the Hebrew University of Jerusalem).

der Nationalhymne, die reichlich fremdartig klingt und sonderbar gegliedert
ist. Japaner sehr andächtig. Unheimliche
Kerle, denen der Staat zugleich Religion
ist. Wetter klarer. Fahrt längs der
Küste zuerst von Sumatra, jetzt längs
des Festlandes. Interessante Brechungs-
erscheinung am Rand des Horizontes durch
Temperatur bzw. Feuchtigkeitsgradienten.
Schiffe scheinen in der Luft zu schwimmen,
ebenso fernes Ufer. Gestern Abend frei-
willige Vorstellung der Japaner. Einer
sang und deklamierte wie ein Kater,
dem man auf den Schwanz tritt, dazu
entlockte er aus einem guitarre-ähn-
lichen Instrument mit einem Brettchen
mit wilder Gebärde von Zeit zu Zeit
einen Ton, der mit dem Sing-Sang
nichts zu thun zu haben schien.
Schlanker vornehmer junger Japaner (Botaniker)
macht erstaunliche Zauberkunststücke,
hauptsächlich mit drei roten Kugeln,
die er verschwinden und wieder auftauchen
lässt. Gestern Navigations-Instrumente gesehen

of the national anthem, which sounds very alien and is strangely structured.[23] Japanese very devout. Weird fellows whose state is at the same time their religion. Weather clearer. Traveling along the coast, first of Sumatra, now along the mainland. Interesting refractive effects at the edge of the horizon. Because of the temperature or humidity gradients, ships seem to float in the air, likewise the far shore. Yesterday evening, spontaneous show by the Japanese. One man sang and wailed like a tomcat whose tail has been stepped on. On top of that, from time to time he coaxed with a wild gesture a tone out of a short-fretted guitar-like instrument that seemed to have nothing to do with the sing-song.[24] Slender, distinguished young Japanese (botanist) performs astonishing magic tricks, mainly with three red balls that he makes disappear and reappear again. Yesterday, saw navigational instruments

und Methoden kennen gelernt, die bei
Ortsbestimmung üblich sind. Kompass ganz
primitiv, mit ausbalang. Trägheitsmoment,
Sextant, Uhr. Geschwindigkeitsmesser, dessen Propeller
langer Leine nachgezogen wird. Morgen früh
Singapore. Frau Ishii entpuppt sich als Tante
von Sakumas junger Braut. Wetter wird
heller.

2. November. 7 Uhr morgens durch enge Passage
zwischen kleinen grünen Inseln hindurch Ankunft
im Hafen von ~~Colombo~~ Singapore. Wir wurden dort von
Wossten erwartet und freundlich begrüsst. Herr
und Frau Montor (er Bruder des Hamburger Scha-
selbst mit schauspielerischer Begabung, sie nach ächter Wienerart, aber in Singapore aufgewachsen)
spielers,) brachten uns in ihr geräumiges Wohnhaus.
Fahrt durch den wundervollen zoologischen Garten
und durch verschiedene Viertel der Stadt bei gutem,
nicht zu warmem Wetter. Ich erfuhr, dass der
unermüdliche Weizmann beschlossen hatte, meine
Reise zionistisch zu verwerten. Im Hause angelangt
musste ich sogleich eine Antwort auf eine Begrüssungs-
Adresse verfassen, die Herr Montor mit einem Freunde
im ~~Auftrag~~ Auftrag des zionistischen Vereins Singapore
verfasst hatte. Die Adresse, welche mir am Nachmittag
feierlich überreicht werden ~~sollte~~, war auf Seide und

and was acquainted with the standard methods for determining position. Compass very primitive, with balanced moment of inertia. Sextant, clock. Speedometer, whose propeller is towed behind on a long line. Tomorrow morning Singapore. Mrs. Ishii turns out to be the aunt of Sakuma's young bride.[25] Weather is brightening up.

2nd November. 7 a.m., through narrow passage between small green islands. Arrival in the harbor of ⟨Colombo⟩ Singapore. Zionists were waiting for us there and gave us a warm welcome. Mr. and Mrs. Montor (*he*, a brother of the Hamburg actor, himself also theatrically talented, *she*, truly Viennese but having grown up in Singapore) brought us to their spacious home.[26] Drive through the wonderful zoological garden and through various parts of town, in not too hot weather. I found out that the indefatigable Weizmann had decided to make use of my voyage for the Zionists' benefit.[27] Having arrived at the house, I immediately had to compose an answer to a welcoming address that Mr. Montor had written with a friend at the behest of the Zionist Association in Singapore. The address, that was to be ceremoniously handed over to me was on silk and

in kostbarem Silberfutteral (Siamesische Reliefarbeit), inhaltlich recht geschickt, nur garzu dick aufgetragen und – wie deren Vater selbst schmunzelnd mitteilte – mit Hilfe von Meyers Convers. Lexikon gebraut. Um 11 Uhr kam ein Journalist dazu, und ich musste Herrn Mentor, seinem Freunde und dem Zeitungsmann unter anderem die Käfer-Geschichte vom sphärischen Raum erzählen. Nach dem prächtigen Mittagessen brachten uns unsere freundlichen Wirte ins Gastzimmer, wo wir unter Mosquito-Käfig schlafen durften. Nebenan war Waschraum mit Closetkübeln und grossem Wascheimer. Else graute es sehr, mir auch ein bisschen wegen dieser ungewohnten Einrichtungen. Gut ausgeruht fuhren wir um 4 Uhr mit Mentors Auto zu Meyer, dem jüdischen Krösus von Singapore. Sein palastartiges Haus mit maurisch anmutenden Hallen liegt auf der Spitze eines Hügels mit Sicht auf Stadt und Meer, unmittelbar darunter eine prunkvolle Synagoge, welche im Wesentlichen eigens für den Verkehr zwischen Krösus und Jehovah von ersterem gebaut war. Krösus ist ein noch recht aufrechter, schlanker willensstarker Greis von 81 Jahren

in a valuable silver case (Siamese ornamental relief), was quite clever in content but far too over the top and—as its creator himself admitted with a grin—concocted with the aid of *Meyer's Convers[ations] Lexikon*.[28] At 11 a.m. a journalist joined us, and I had to tell Mr. Montor, his friend, and the newspaper man, the beetle story about spherical space, among other things. After the splendid midday meal our friendly hosts took us to a guestroom where we were permitted to sleep under a mosquito cage. Next door was a lavatory with chamber pots and a big washtub. Elsa was horrified, so was I a little, in light of such unaccustomed installations. Well-rested, we drove in Montor's car at 4 p.m. to Meyer, the Jewish Croesus of Singapore.[29] His palatial house with Moorish-like halls is situated on the peak of a hill with a view of city and sea; directly below it, a sumptuous synagogue,[30] essentially solely for dealings between Croesus and Jehovah, built by the former. Croesus is a still quite upright, slim, strong-willed geriatric, 81 years old.

Graues Spitzbärtchen, schmales rötliches Gesicht, jüdische gebogene schmale Nase, kluge, etwas schlaue Augen, ein schwarzes Käppchen auf der wohlgewölbten Stirn. Ähnlich wie Lorentz, nur dass dessen glänzende, wohlwollende Augen durch vorsichtig schlaue ersetzt sind und der Gesichtsausdruck mehr von schematischer Ordnung und Arbeit als – wie bei Lorentz – von Menschenliebe und Gemeinsinn spricht. Es war die Festung, welche ich nach Weizmanns Absicht zugunsten der Jerusalemer Universität einnehmen sollte. Seine Tochter, schmales blasses adliges Gesicht ist eine der feinsten jüdischen Frauengestalten welche ich gesehen habe. Wenn man sie ansieht, ist man versucht, den Scherz vom „ältesten Adel" ernst zu nehmen.

Als wir nun da oben ankamen, gings querst aus Photographieren. Kuns neben mir, herum Familie und viel jüdische Ehepaare, Gruppenbild. Nach diesem enorm wichtigen Akt gings in eine orientalische grosse Erfrischungshalle. Malaysche Kapelle spielte Wienerisches und Negerisches in europäisch-schmalziger Kaffee–Haus–Manier. Ich sass mit Kuns, und dem – Erzbischofsrebst

Gray goatee, narrow ruddy face, thin Jewish hooked nose, clever, somewhat sly eyes, a black skull cap on his well-arched brow. Similar to Lorentz, only that his shining benevolent eyes are replaced by cautiously shrewd ones and his demeanor expresses schematic order and work rather than—with Lorentz—human kindness and communal spirit.[31] This was the fortress that, according to Weizmann's designs, I was to conquer for the benefit of the Jerusalem University. His daughter, narrow ⟨black⟩ pale noble face, is one of the finest Jewish female figures I have ever seen. ⟨She proves⟩ Looking at her, one is tempted to take seriously the joke about the "oldest aristocracy."[32]

When we arrived up there, first came the business of picture taking. Croesus next to me, surrounded by his family and many Jewish couples, a group photo.[33] After this enormously important act we filed into a spacious oriental refreshment hall.[34] Malayan band played Viennese and Negro music in European schmaltzy coffeehouse manner. I sat with Croesus and the—archbishop (together

9. The Einsteins with members of the Singapore Jewish community at the residence of philanthropist Manasseh Meyer, 2 November 1922. Front row: Alfred and Anna Montor, Einstein, Manasseh Meyer, Elsa Einstein, Mr. and Mrs. Weil, and Rosa Frankel. Back row: unidentified man, Mozelle Nissim, Julian Frankel, Charles R. Ginsburg, Tila Frankel, Victor Clumeck, Marie Frankel-Clumeck, and Abraham Frankel (photo appears in *The Jews of Singapore* by Joan Bieder. Permission courtesy of Lisa Ginsburg).

(Gemahlin) einem ausgekochten, nur englisch
sprechenden schlanken grossnäsigen
englischen Nobile, der mit Kiams' Geld nicht
ohne Glück liebäugelt, ohne dessen Seele
im Geringsten für sich zu beanspruchen.
verzweifelte Sprach-Kalamität mit wohlschmecken-
den Kuchen. Nun brachte uns Mentor (Kiams
und noch) auf zwei am Saalende vorgesehene
erhöhte Stühle mit Rednertisch, setzte sich
daneben, liess mit überzeugender Stimme seine
Adresse vom Stapel, meine wohlübersetzte
Antwort in der sicheren Tasche. Ich antwortete frei, wobei er Notizen simulierte, um
dann darauf meine morgens redigierte und
mit ihm und seinem Freunde improvisierte
übersetzte Antwort zu verlesen, die er als Über-
setzung meiner freien Ansprache ausgab,
der Teufelsmeier. Dann mündliches Händeschütteln,
an Amerika erinnernd. Echte Herzlichkeit bei
den Juden überall. Als ich schon ganz seekrank
davon war und die Sonne sich verkrochen hatte,
fuhren wir schnell nachhause, wo ich einen
kleinen Stoss Albümer mit Autographen ver-
sehen musste. Dann fuhren wir durch das Chinesenviertel (ungeheures Gewusel, aber die Zeit reichte

with spouse), a crafty, exclusively English-speaking, slim, big-nosed English nobleman, who, not in vain, ogles Croesus's money, without in the least claiming his soul.[35] Hopeless linguistic calamity with tasty cake. Then Montor brought us (Croesus and me) to two raised seats with lectern set up at the end of the hall, sat down beside us, delivered his address with a convincing voice, with my well-translated reply safely in his pocket. I spoke freely, while he simulated taking notes, in order to be able thereafter to read out my answer, composed that morning and translated by him and his friend, which he gave off as an improvised translation of my impromptu address, the slick operator. Then endless handshaking, reminiscent of America. Genuine cordiality among Jews everywhere. Already quite seasick from the whole affair, and the sun had already set, we quickly drove home, where I had to provide autographs for a small pile of albums. Then we drove through the Chinese quarter (tremendous bustle but there wasn't enough time

nicht zum Sehen sondern nur zum Riechen)
zu Krösus' Abendessen, einer pompösen Mahl
zeit in offener Halle für etwa 80 Menschen. Die
Mahlzeit war deftig und ohne Ende. Ich musste
endlich aufstehen, weil ich Speisen nicht einmal
mehr sehen, geschweige denn essen konnte.
Nun kam die bewusste Kapelle wieder und
negerte lustig drauf los, der Alte und alles
tanzte. Dies verschmähte sogar Krösus nicht,
nachdem er gezeigt hatte, dass sein 80-jähriger
Magen noch gewaltig leistungsfähig war.
Endlich fand das wohlgeplante, von Weizmann
veranlasste Attentat auf Krösus statt, von dem
ich trotz vieler Bemühungen nicht weiss, ob
eins meiner Geschosse die dicke Oberhaut
Krösus' hat durchdringen können. Dann
fuhren wir heim und schlüpften (nach Absol-
vierung noch einiger Albümen) in den wohlverrammten
Mosquito-Käfig. Nachts ungeheuer heftiger
Tropen-Regenguss mit Gewitter. Ich stopfte
schnell alle Fensterläden zu, was aber doch
nur unvollständig gegen das viele Wasser
half. Die Temperatur war während des ganzen
Aufenthaltes nicht besonders hoch, aber die

to look, only to smell) to Croesus's dinner, a pompous repast in an open hall for about 80 people.[36] The meal was hearty and endless. I had to finally get up because I couldn't even look at food anymore, let alone eat it. Then the purposeful band returned and merrily played Negro music,[37] ⟨The geriatric⟩ and everyone danced. Not even Croesus spurned doing so, after having shown how mightily fit his 80-year-old stomach still was. At last the well-planned assassination attempt instigated by Weizmann on Croesus took place (about contribution to Jerusalem University); I do not know whether, despite much effort on my part, one of my missiles penetrated Croesus's thick epidermis.[38] Then we drove home and (after finishing off a few more albums) slipped into the well-deserved mosquito cage. At night, enormously hefty tropical downpour with thunderstorm. I quickly secured all the window shutters, which nevertheless only partially helped against so much water. Throughout the stay, the temperature was not particularly elevated, but the grosse Feuchtigkeit erinnert ans Treibhaus. Hat etwas Morgenländisch-Berauschendes.

3. November. Nach dem Frühstück wundervolle Auto-fahrt über Gummi-Plantagen-Hügel nach dem Hafen. Wundervolle Vegetation, fröhlich aussehende chinesische Villen. Aussicht über Meer mit Inselchen. Noch ein schönes Häuflein gebräunter Inder kam aufs Schiff. Erst gegen 11 Uhr gings ab in malerischer Fahrt zwischen grünen Inseln.

Die Chinesen vermögen jedes andere Volk zu verdrängen durch Fleiss, Anspruchslosigkeit, Kinderreichtum. Singapore ist fast ganz in ihren Händen. Geniessen als Kaufleute grosse Achtung, weit mehr als Japaner, die als unzuverlässig gelten. Es mag schwer sein sie psychisch zu verstehen, ich habe Scheu davor, seitdem mir der japanische Gesang so ganz unverständlich geblieben ist. Gestern hörte ich wieder einen vor sich hin singen, dass es mir schwindlig davon wurde.

great humidity reminds one of a greenhouse. There is something orientally intoxicating about it.

3rd November. After breakfast, wonderful drive over rubber plantation hills to the harbor. Splendid vegetation; cheerful-looking Chinese villas. View over sea with islets. Another fine bunch of faithful Jews come onboard. Cast off only around 11 a.m. in a scenic trip past green islands.

The Chinese may well supplant every other nation through their diligence, frugality, and abundance of offspring. Singapore is almost completely in their hands. They enjoy great respect as merchants, far more than the Japanese, who are deemed unreliable. It may be hard to understand them psychologically, I hesitate to try ever since the Japanese singing remained so entirely incomprehensible to me. Yesterday I heard another one singing away again to the point of making me dizzy.

7. XI. In der Zwischenzeit Regenwetter bei
Treibhausluft. Gesellschaft auf Dampfer in
Singapore um zwei alte gemütliche Schweizer-
Offiziere und jungen deutschen Kaufmann
bereichert. Vom Abend des 5 bis Abend des 6. Taifun
mit riesigen Wellen, Wind und viel heftigem
Regen. Das Schiff tanzte gewaltig. Wunderbares
Schauspiel an Schiffspitze. Viel fliegende Fische,
vom Dampfer aufgescheucht; gewaltige Vertikal-
bewegung. Heute zeigt das Meer noch Remi-
niszenzen in seiner Bewegung. Ermüdung durch
ewiges Balancieren. Frauen viel mehr seekrank
als Männer.

10. Am 9. Morgens kamen wir in Hongkong
an. Es ist die schönste Landschaft, die ich
auf der ganzen Reise bisher gesehen habe. Gebirgige,
langgestreckte Insel nahe dem ebenfalls
gebirgigen Festlande. Zwischen beiden der
Hafen. Viele kleine steile Inselchen. Das Ganze
wie eine halb ersoffene Gegend aus den
Voralpen. Die Stadt terrassenartig am mässig
geneigten Fuss eines etwa 500 m hohen Berges
gelegen. Luft angenehm kühl.

   Den Empfang der jüdischen Gemeinde dankbar

7th Nov. In the interim, rainy weather with greenhouse temperatures. In Singapore, company on the steamer enriched by two old jovial Swiss officers and a young German businessman. From evening of 5th to evening of 6th, typhoon with huge waves, wind, and much hefty rain. The ship danced about mightily. Wonderful show at the prow. Many flying fish startled by the steamer; powerful vertical motion. Today the sea still presents vestiges in its motion. Fatigue from eternal balancing act. Women far more seasick than men.

10th. On the morning of the 9th we arrived in Hong Kong. It is the prettiest landscape I've seen up to now on the entire voyage. Mountainous, stretched-out island next to the similarly mountainous mainland. Between the two, the harbor. Many small steep islets. The whole thing like a half-drowned area from the Alpine foothills. The city lies terrace-like at the foot of a gently sloping, roughly 500 meters high mountain.[39] Air pleasantly cool.

Gratefully declined the reception by the Jewish community.[40]

abgelehnt. Aber zwei jüdische Kaufleute ver-
brachten den ganzen Tag mit uns. Vormittags
Rundfahrt um Insel im Automobil angetreten.
Blicke auf Meer, fjordähnliche Meeresbuchten
und Berghalden von unerschöpflicher
Mannigfaltigkeit und Pracht. Unterwegs
assen wir zu Mittag in amerikanisch
anmutendem Luxushotel, wo unsere
beiden Lehrer sich nicht nur angeregt über
Land und Wissenschaft mit mir unterhielten,
sondern auch eine grosse Zuneigung zu
irdischen Genüssen offenbarten. Bei der
Heimfahrt sahen wir ein aus Barken be-
stehendes chinesisches Fischerdorf, eine
sehr fröhlich anmutende chinesische Beerdigung und -geplagte Menschen, Männer
und Weiber, die für 5 Cent täglich Steine klopfen
und Steine tragen müssen. So werden die
Chinesen für ihre Fruchtbarkeit von der
fühllosen Wirtschaftsmaschine hart ge-
straft. Ich glaube, sie merken es kaum
in ihrer Stumpfheit, aber traurig zu sehen
ist es. Sie sollen übrigens vor einiger Zeit
mit merkwürdig guter Organisation einen erfolg-

But two Jewish businessmen spent the entire day with us.[41] In the morning, started tour around the island by automobile. Views onto sea, fjord-like bays, and mountain slopes of inexhaustible variety and magnificence. Along the way we ate lunch in a seemingly American-style luxury hotel,[42] where both our guides not only conversed with me animatedly about the country and science but also revealed a great affinity for worldly pleasures. On the homeward trip we saw a Chinese fishing village composed of sailing barks, a seemingly very cheery Chinese funeral and—stricken people, men and women, who beat stones daily and must heave them for five cents a day. In this way, the Chinese are severely punished for their fecundity by the insensitive economic machine. I think they hardly notice it in their obtuseness, but it is sad to see. Incidentally, they supposedly carried out a successful

lohnstreik erfolgreich durchgeführt haben.
Nachmittags besuchten wir das jüdische Klubhaus, das in einem üppigen Garten in einiger
Höhe gelegen ist und einen prächtigen
Ausblick auf Stadt und Hafen hat. Es sind
nur 120 Juden dort, meist arabische, deren
Religiosität noch mehr zur Form erstarrt zu
sein scheint, als dies bei unseren russisch-
europäischen der Fall ist. Im Klubhaus gesellten
sich die zwei Frauen zu uns, die Frau eines unserer
Gastgeber und deren Schwester. Ich bin nun
überzeugt, dass die jüdische Rasse sich ziemlich rein erhalten hat in den letzten 1500 Jahren,
da die Juden aus den Euphrat-Tigris-Ländern
den unsrigen sehr ähnlich sind. Das Gefühl der
Zusammen-Gehörigkeit ist auch recht lebhaft.
Wir fuhren noch alle zusammen auf den Gipfel
des Berges, an dessen Fuss die Stadt liegt (mit
Drahtseilbahn; Chinesen und Europäer getrennt).
Oben grandiose Aussicht auf Hafen, Inselgebirge und
Meer. Der Anblick mit den vielen kleinen
Inselchen, die steil aus dem Meere anragen,
erinnert an das Nebelmeer in den Voralpen.
Abends kam plötzlich Sturm, da wir auf der

wage strike with remarkably good organization a while ago.[43] In the afternoon we visited the Jewish clubhouse, which is set in a lush garden at quite a high elevation and has a magnificent view of city and harbor.[44] There are supposedly only 120 Jews there, mostly Arab, whose religiosity seems to have ossified more into formality than is the case with our Russo-Europeans. In the clubhouse two women socialized with us, the wife of one of our hosts and her sister. I am now convinced that the Jewish race has maintained its purity in the last 1,500 years, as the Jews from the lands of the Euphrates and the Tigris are very similar to ours. The feeling of belonging together is also quite strong. We also all drove together to the summit of the mountain, at the foot of which the city lies (by cable railway; Chinese and Europeans segregated). At the top, grandiose view onto harbor, island mountains, and sea. The sight of the many small islets looming steeply out of the sea reminds one of the sea of mist in the Alpine foothills. In the evening there was a sudden storm that

Strasse meinen Hut entriss, sodass ich
mit allen Kräften nachlaufen musste,
um ihn wieder zu bekommen.

Heute Morgen besuchte ich mit Ilse das
Chinesenviertel auf der Festlandseite. Fleissiges,
dreckiges, stumpfes Volk. Häuser sehr scha-
blonenhaft, bienenzellenartig ~~eingegittert~~ voneinander,
alles zusammengebaut und eintönig. Hinter
dem Hafen lauter Essläden, vor denen
Chinesen auf Bänken bei der Mahlzeit
nicht sitzen sondern hocken wie Europäer,
wenn sie im grünen Walde ihre Notdurft verrichten.
Es geht still und gesittet bei allem zu. Schon
die Kinder sind temperamentlos und sehen
stumpf aus. Es wäre doch schade, wenn diese
Chinesen alle andern Rassen verdrängten. Für
unsereinen ist schon der Gedanke daran
unsäglich langweilig. Gestern Abend besuchten
mich noch drei Portugiesen (Mittelschullehrer), die von den Chinesen
behaupteten, dass Chinesen nicht zu logischem
Denken erzogen werden könnten, insbesondere
gar keine Begabung für Mathematik hätten.
Mir fiel der geringe Unterschied zwischen Männern
und Weibern auf; ich begreife nicht, was für —

snatched away my hat into the street, so I had to run after it with all my might to retrieve it.

This morning I visited the Chinese quarter on the mainland side with Elsa. Industrious, filthy, lethargic people. Houses very formulaic, balconies like beehive-cells, everything built close together and monotonous. Behind the harbor, nothing but eateries in front of which Chinese don't sit on benches while eating but squat like Europeans do when they relieve themselves out in the leafy woods. All this occurs quietly and demurely. Even the children are spiritless and look lethargic. It would be a pity if these Chinese supplant all other races. For the likes of us the mere thought is unspeakably dreary. Yesterday evening three Portuguese middle-school teachers visited me, who claimed that the Chinese are incapable of being trained to think logically and that they specifically have no talent for mathematics. I noticed how little difference there is between men and women; I don't understand what

eine Art Reiz der Chinesinnen die zugehörigen
Männer so fatal begeistert, dass sie sich
gegen den formidabeln Kindersegen so schlecht
zu wehren vermögen. Um 11 Uhr fuhr die
Kotano Maru ab, durch das glänzende grüne
Meer zwischen grünen Inselbergen, die
entzückend waren durch Form und Farbe,
aber kahl, d. h. ohne Baumwuchs. Die jetzige
üppige Flora auf Hongkong soll ganz von den Engländern
angelegt sein. Diese verstehen das Regieren
bewundernswürdig. Die Polizei wird
durch importierte schwarze Inder von wun-
derbarem Wuchs besorgt, niemals werden Chinesen
verwandt. Für letztere haben die Engländer eine
richtige Universität errichtet, um die in ihrer
Lebenshaltung emporgestiegenen Chinesen an
sich zu fesseln. Wer macht ihnen das nach?
Arme europäische Kontinentale, ihr versteht
nicht, nationalen Gegenbewegungen durch Tolerierung
die Gefährlichkeit zu nehmen.
11. Nachts wunderbares Meeresleuchten. Die Kämme
der Meereswogen leuchteten bläulich soweit man
sehen konnte.
14. Am 13. etwa 10 Uhr morgens Ankunft in Schanghai

kind of fatal attraction Chinese women possess that enthralls the corresponding men to such an extent that they are incapable of defending themselves against the formidable blessing of offspring. At 11 o'clock the [S.S.] *Kitano Maru* left through the shining green sea between green island mountains that were delightful in shape and color but bald, i.e., without tree growth. The current lush flora on Hong Kong is supposed to have all been planted by the English. They have an admirable understanding of governance. The policing is carried out by imported black Indians of tremendous stature, Chinese are never used. For the latter the English have established a proper university in order to bind those Chinese who have prospered closer to them.[45] Who can emulate them in that? Poor Continental Europeans, you don't understand how to take the bite out of nationalistic opposition movements by means of tolerance.

11th. At night, wonderful marine phosphorescence. The crests of the sea's waves glowed bluish as far as one could see.

14th. On the 13th, about 10 a.m., arrival in Shanghai.

## 14

Fahrt an flachen, malerisch gelbgrün beleuchteten Ufern entlang flussaufwärts. Abschied von den zwei Schweizeroffizieren, deren einer aus Bern mir in liebenswürdigster Weise mein Pfeifchen geflickt hat und von chauvinistischen, aber sonst gutherzigen jungen Deutschen, ehemaligem Offizier. In Schanghai von Inayuki und Frau, unseren liebenswürdigen Begleitern Schanghai – Kobe, vom Deutschen Konsul, Herrn und Frau Pfister auf dem Schiff begrüsst. Zuerst Journalisten, ein ansehnliches Häufchen japanische und amerikanische, die ihre gewohnten Fragen stellten. Dann mit Inayukis und zwei Chinesen (ein Journalist und Sekretär christlicher Chinesenverbände) in chinesisches Restaurant geführt. Während des Essens sahen wir zum Fenster hinaus eine geräuschvolle, farbige chinesische Beerdigung — eine für unseren Geschmack barbarisch, fast drollig anmutende Angelegenheit. Das Essen höchst raffiniert, schier endlos. Man greift ungesetzt mit Stäbchen aus gemeinsamen Schüsselchen, die in grosser Anzahl auf dem Tisch stehen. Mein Inneres reagierte recht temperamentvoll, so dass es höchste Zeit war, als ich gegen 5 Uhr im Hafen (buchstäblich

Travel upriver along flat, picturesque, yellowish-green illuminated shores.[46] Farewell to the two Swiss officers, the one from Bern who so kindly mended my little pipe, as well as the chauvinistic but otherwise kindhearted young German former officer. In Shanghai, greeted on the ship by Inagaki and spouse, our dear escorts Shanghai-Kobe,[47] and by the German consul,[48] [and] Mr. and Mrs. Pfister.[49] First, journalists, a respectable bunch of Japanese and American ones who posed their usual questions. Then, with Inagakis and two Chinese (a journalist and secretary of Christian Chinese associations), led to a Chinese restaurant.[50] During the meal we observed through the window a noisy, colorful Chinese funeral—a ⟨somewhat⟩—for our taste—barbaric, almost seemingly comical affair. The food, extremely sophisticated, interminable. One constantly fishes with sticks from common little bowls set out on the table in great numbers. My innards reacted quite temperamentally, so it was high time when I landed around 5 p.m. in the safe haven (to be understood

zu verstehen) des freundlichen Ehepaars Pfister
landete. Nach dem Essen bei herrlichem Wetter
Spaziergang durch das Chinesenviertel. Strassen
immer enger, wimmelnd von Fussgängern, Kuli-
Personenwägelchen, starrend von Dreck aller Art,
in der Luft ein Gestank von nicht endendem
mannigfaltigem Wechsel. Eindruck von grässlichem Existenzkampf sanft und meist stumpf
aussehender meist vernachlässigter Menschen. Nach
der Strasse lauter offene Werkstätten und Läden, grosses
Geräusch, aber nirgends Streit. Wir besuchten Theater,
in jedem Stock besondere Vorstellung von Komikern.
Publikum stets dankbar, sehr ergötzlich, verschiedenstes Volk mit
kleinen Kindern. Überall respektables Dreck.
Drin und draussen in dem schrecklichen Gewimmel
ziemlich frohe Gesichter. Sogar die zu Pferdarbeit
Degradierten machen mir den Eindruck bewussten
Schmerzes. Merkwürdiges Herdenvolk, oft respektabel,
Bäuchlein, immer gute Nerven, oft mehr Automaten
als Menschen ähnelnd. Manchmal Neugierde mit
Grinsen. Bei Europäerbesuch wie war drolliges gegen-
seitiges Anglotzen – Else besonders eindrucksvoll
mit aggressiv anmutender Lorgnette.
Dann Fahrt zu Pfisters geräumigem Landhaus mit schon ge-

literally) of the friendly Pfisters. After the meal, stroll through the Chinese quarter in splendid weather. Streets becoming ever narrower, swarming with pedestrians, rickshaws, caked with dirt of every kind, in the air there is a stench of never-ending manifold variety. Impression of ghastly fight for survival by meek and mostly lethargic-looking, mostly neglected people. Beyond the street, loud open workshops and shops, great racket but nowhere any quarreling. We visited a theater, on every floor a separate performance by comedians.[51] Audience always appreciative, very delighted, the most diverse people with small children. Decent filth everywhere. Inside and outside in the terrific bustle, quite happy faces. Even those reduced to working like horses never give the impression of conscious suffering. A peculiar herd-like nation, often a respectable paunch, always sound nerves, often resembling automatons more than humans. Sometimes curiosity with grinning. With European visitors like us, comical mutual staring—Else particularly imposing with seemingly aggressive lorgnette.

Then, drive to Pfisters' spacious country house with already

rühmten rettenden Hafen. Gemütlicher Thee.
Dann kam eine Deputation von etwa 8 jüdischen
Honoratioren mit dem würdigen Rabbi und
recht schwieriger Verständigung. Dann Fahrt
mit Inagakis durch dunkle Gassen zu
reichem chinesischem Maler zu chinesischem
Abendmahl. Haus außen dunkel mit
kahler hoher Mauer. Innen festlich beleuchtete Hallen um einen mit malerischem
Teich und Garten ausgestatteten romantischen
Hof. Die Hallen mit prächtigen ächt chinesischen
Bildern des Hausherrn geschmückt und
mit liebevoll gesammelten alten Kunstgegen-
ständen. Vor dem Essen ganze Tischgesellschaft
bestehend aus dem Hausherrn, uns, Inagakis Pfisters,
einem deutsch sprechenden chinesischen, dem Hausherrn verwandtes Ehepaar mit zutraulichem,
deutsch und chinesisch allerliebst deklamierendem
etwa 10 jährigem hübschen Töchterchen, dem Rektor
der Schanghaier Universität und ein paar Lehrern
dieser Anstalt. Endloses, ungeheuer raffiniertes
Fressen, einem Europäer unvorstellbare, geradezu
lasterhafte Schlemmerei mit schmatzigen, von
Inagaki hin und her übersetzten Reden, hievon eine von mir.

praised safe haven. Cozy teatime.[52] Then came a deputation of about eight Jewish dignitaries with venerable rabbi and very difficult communication.[53] Then, drive with Inagakis through dark alleyways to a wealthy Chinese painter for a Chinese supper. House dark outside with bleak high wall. Inside, festively lit halls around a romantic courtyard furnished with a picturesque pond and garden. The halls adorned with the host's magnificent authentic Chinese pictures and with lovingly collected old art objects. Prior to the meal, the entire dinner party including the host, us, Inagakis, Pfisters, a German-speaking Chinese, a couple related to the host with a trusting, pretty little daughter, about 10 years old, most sweetly reciting in German and Chinese, the rector of Shanghai University and a few teachers from that institution.[54] Endless, extremely fancy fare, inconceivable for a European, downright sinful indulgence with schmaltzy speeches translated back and forth by Inagaki, one of them by me.[55]

10. Einstein and Elsa attending dinner at residence of prominent Shanghai painter and entrepreneur Yiting Wang, 13 November 1922. Front row: Shu Zhang(?), Tony Inagaki, Elsa Einstein, Huide Ying, Einstein, Morikatsu Inagaki, Yiting Wang, and Youren Yu (courtesy of the Leo Baeck Institute, New York).

Der Hausherr hatte ungemein feines Gesicht, Haldane ähnliche. An der Wand hing ein wundervolles, lapidares Selbstbildnis von ihm. Die Mutter des deklamierenden Töchterchens spielte die Hausfrau und führte recht drollig und geschickt auf Deutsch die Unterhaltung. Um 9½ Uhr Abfahrt mit Inagakis in den japanischen Klub, wo wir von etwa hundert meist jungen Japanern in angenehm formloser, schlichter und heiterer Weise willkommen geheissen wurden. Zwanglose Begrüssung und Beantwortung derselben, übersetzt von Inagaki. Dann Rückkehr auf das Schiff. Dort noch Besuch von interessantem englischen Ingenieur. Endlich Bett.

Heute nach Frühstücks Autofahrt nach interessantem ... buddhistischen Tempel mit prächtigem chinesischen Turm. Nebenan höchst amüsantes Dörfchen, ganz chinesisch mit ganz engen Gässchen und nach vorn offenen Häuschen, überall kleinen Laden oder Werkstätte darin. Gegenseitiges Anglotzen noch persönlicher als in der Stadt. Kinder schwanken zwischen Neugierde und Furcht. Fast durchweg fröhlicher Eindruck nebst Dreck und Gestank. Ich werde oft und gerne dran denken. Den Tempel besehen wir uns genau

The host had an exquisitely fine face, similar to Haldane.[56] On the wall hung a wonderful lapidary self-portrait of him. The mother of the little reciting daughter played hostess and led the conversation quite amusingly and ably in German. At 9:30 departure with the Inagakis to the Japanese Club, where we were greeted by about a hundred mostly young Japanese in a pleasantly informal, modest and cheerful way.[57] Same casual welcome and response, translated by Inagaki. Then return to the ship. There, another visit by an interesting and sympathetic English engineer. Finally, bed.

Today after breakfast, trip by car to interesting temple with many courtyards and a gorgeous Chinese tower, presently being used as barracks. Next door, a highly amusing village, entirely Chinese with very narrow alleyways and little houses opening toward the front, small shops or workshops everywhere. Reciprocal staring even more comical than in the city. Children teeter between curiosity and fear. Almost throughout, cheerful impression along with filth and stench; I shall remember it often and with pleasure. We had a close look at the temple.

Die benachbarten Menschen scheinen gegen seine
Schönheit stumpf zu sein. Architektur und
innere Ausstattung (überlebensgrosse Budda's
und andere Figuren) wirken merkwürdig
zusammen zu grossem künstlerischen
Gesamteindruck. Hoheit des buddhistischen
Gedankens umrankt von ~~Gestalten abstrusen~~ barockartig umrankenden
Aberglaubens (halb symbolisch).
    Nachmittags 3 Uhr Abreise.
16. und 17. Fahrt durch japanische Meerengen mit
unzähligen grünen Inselchen. Wunderbare
stets wechselnde Nordlandschaften. 17. Nachmittags Ankunft in Kobe. Von Nagaoka Ishiwara
Kuroka (deutschem Kunst und deutschem Verein)
empfangen. Grosser Trubel. Auf dem Schiff
Haufen Journalisten. Halbe Stunde Interview
im Salon, kurzes Verschnaufen im Hotel nahe
Landungsstelle. Abends 2 Stunden Bahnfahrt
mit den Professoren. Leichte Wagen. Publikum sitzt
in zwei langen Reihen ~~gg~~ längs Fenster. In Kioto
magisch beleuchtete Strassen, niedliche Häuschen.
Autofahrt ins Hotel. Grosser Eindruck. Graziöse
Leutchen trippeln klapp klapp auf den Strassen.
Hotel grosser Holzbau. Gemeinsames Essen, zierliche

The neighboring people seem to be indifferent toward its beauty. Architecture and interior decor (larger than life-sized Buddhas and other figures) create a strange effect together to form a great artistic total impression. Lofty Buddhist thought amidst seemingly baroque-like figures of abstruse superstition (half symbolic).

In the afternoon, 3 p.m. departure.

16th and 17th. Passage through Japanese straits with countless green islets. Delightful, ever-changing fjord landscapes. 17th, in the afternoon, arrival in Kobe. Welcomed by Nagaoka, Ishiwara, Kuwaki, Mr. Nagaoka with dainty wife, German consul and German Club, Zionists.[58] Great hubbub. Masses of journalists on board the ship. Half-hour interview in the saloon.[59] Disembarkation with huge crowds. Brief breather in the hotel next to the wharf.[60] In the evening, two hour train ride with the professors. Simple carriages. Passengers sit in two long rows along the windows. In Kyoto, magically lit streets, cute little houses. Drive to hotel somewhat high up.[61] The city below, like a sea of lights. Grand impression. Graceful little people patter about, click-clack, on the streets. Hotel, a large wooden structure. Communal meal; dainty

Frauen - Bedienung in kleinem Separatzimmer.
Japaner schlicht, fein, überhaupt sehr sympathisch.
Abends wissenschaftliche Unterhaltung. Alles
zusammen höchst anstrengend.

18. Morgens Auto - Rundfahrt durch Kioto. Tempel.
Grosse Gärten, alter Palast, mit Mauer und
Wassergräben umgeben, wundervolle altjapanische
Architektur (Abart der chinesischen, leichter, weniger
barock) Auf Strassen allerliebste Schulkinder.
Von 9 Uhr Morgens bis 7 Uhr Abends Eisenbahnfahrt im Aussichtswagen bei wolkenlosem
Himmel bis Tokio. Fahrt an Seen, Meeresbuchten.
Über Gebirgspass Fushiama, schneebedeckt
strahlt weithin über das Land. Unvergesslich-
licher Sonnenuntergang nahe beim Fushi. Pracht-
volle Silhouetten der Wälder und Hügel. Dörfer
lieblich und sauber. Schöne Schulen. Land sorg-
fältig bebaut. Neue Sonnenuntergang. Journalisten
im Zug. Gewohnte blöde Fragen wie immer.
Ankunft in Tokio! Gewimmel von Volk und
Photographen mit Blitzlichtern. Wurden völlig
geblendet von unzähligen Magnesium - Blitzen.
Kurze Ruhe Zufluchtsaufenthalt in Halle im
Bahnhofhotel. Empfang von Akademie und

waitresses in small separate room. Japanese unostentatious, decent, altogether very appealing. In the evening, scientific conversation. All in all, highly strenuous.

18th. In the morning, automobile tour around Kyoto. Temple. Large gardens, old palace surrounded by wall and moats, wonderful ancient Japanese architecture (derived from Chinese, simpler, less baroque).[62] On streets the dearest of schoolchildren. From 9 a.m. until 7 p.m., train ride to Tokyo in observation car under cloudless skies. Trip along lakes and sea coves. Over the Fujiyama mountain pass, snowcapped, gleams far out over the land.[63] Unmatched sunset near [Mt.] Fuji. Magnificent silhouettes of woods and hills. Villages quaint and clean. Beautiful schools. Land carefully cultivated. After sunset, journalists in the train. Usual daft questions as always. Arrival in Tokyo! Throngs of people and photographers with flashbulbs. Were completely blinded by countless magnesium flashes.[64] Brief refuge in lobby in train station hotel.[65] Reception by academy, and

Deutschen und Vereinsdeputation. Ankunft
ins Hotel ganz erschöpft zwischen riesigen
Blumenkränzen und Bouquets. Lebendiges
Begrüßung.

19. Von 1½ – 4 und 5 – 7 öffentlicher Vortrag
in Universitätssaal in kleinen Portien,
übersetzt von Ishwara. Dieser malerisch
in japanischer Kleidung sieht aus wie Mittel-
ding von Büsser und Priester.

20. Nagaoka holte mich ab zur Akademie-Essen
in botanischen Garten. Die Akademiker
waren sehr herzlich. Nagaoka holte uns
ab und brachte uns heim. Es wurde uns
nach Vorlesung eine Adresse überreicht,
für die ich kurz dankte. Abendessen im
Hotel zusammen mit Yamamoto und
Kaizosha-Angestellten. Dann japanisches Theater
mit Gesang und Tanz. Frauenrollen von
Männern gegeben. Publikum sitzt am Boden
in lauter kleinen Logen mit Kind und Kegel
& beteiligt sich lebhaft. Zugänge zur Bühne
durch Parterre, welche Durchgänge ebenfalls
zur Bühne gehören. Rollen stark stilisiert. Chor
aus 3 Männern singen unaufgesetzt ähnlich

11. Einstein and Elsa at the Koishikawa Botanical Gardens with members of the Imperial Academy of Japan, Tokyo, 20 November 1922. Center of front row: Tetsujiro Inoue, Therese Nagai-Schumacher, Einstein, Nobushige Hozumi, Elsa Einstein, and Hantaro Nagaoka (courtesy of the Japan Academy).

Germans, and association deputations.[66] Arrival at hotel, completely exhausted among gigantic floral wreaths and bouquets. Still to come: visit by Berliners[67] and burial alive.

19th. From 1:30–4 and 5–7, public lecture in university auditorium in small segments, translated by Ishiwara.[68] The latter quaint in Japanese clothing. Looks like a cross between penitent and priest.

20th. Nagaoka picked us up for the academy dinner in the botanical gardens. The academicians were very cordial. Nagaoka collected us and brought us home. An address was read out and then handed to me, for which I gave brief thanks.[69] Supper in the hotel together with Yamamoto and Kaizosha employees.[70] Then Japanese theater with song and dance.[71] Female roles performed by men. Audience sits on the floor in separate little boxes with whole kit and caboodle & participates in an animated manner. Access to stage through ground floor, the passageways of which also belong to the stage. Roles strongly stylized. Chorus of three men sing incessantly, similar to

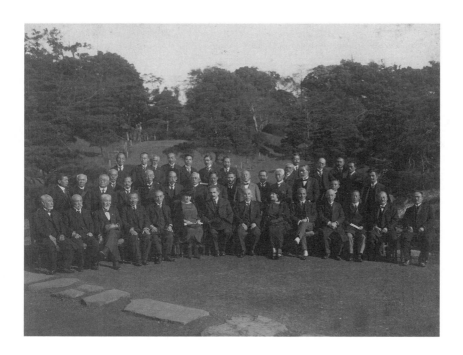

Puvertun auf der Messe. Orchester in einer Art Käfig hinter der Szene. Malerische Szenerie. Musik gibt Rhythmus und Gefühlston, Seyley...? vergleichbar, ohne symphonische Logik und Geschlossenheit. Schauspieler pathetisch und auf malerische Effekte bedacht. Dann mit Inagaki und Ehepaar Yamamoto Bummel durch Ginzastrasse mit allerlei hübschem Tand für Kinder und Erwachsene. Helle Beleuchtung überall, aber wenig Volk wegen verhältnismässig grosser Kälte. Grosse Geschäfts- und Verkehrsstrasse um 10 Uhr so gut wie leer. Wir gingen in reizendes kleines Restaurant und plauderten. Dann heim, mit Obst und Zigarren, von dem ... Y. geschenkt. 21. Chrysanthemen-Fest im kaiserlichen Schlossgarten. Grosse Schwierigkeit in der Beschaffung eines passenden Gehrocks nebst Zylinder. Ersteren von unbekannter Seite durch Herrn Bärwald, der ihn eigens brachte, letzteren durch Herrn Yamamoto; viel zu klein, sodass ich ihn den ganzen Nach- Mittag in der Hand tragen musste. Wir waren mit den ausländischen Diplomaten,

priests at mass. Orchestra in a kind of cage backstage. Picturesque scenery. Music provides rhythm and emotional expression, comparable to bird chirping, lacking symphonic logic and unity. Actors display pathos and intent on pictorial effects. Then, with Inagaki and the Yamamotos, saunter through shopping street with booths of all sorts of pretty trinkets for children and adults. Bright lighting everywhere, but few people, owing to relatively cold weather. Large business thoroughfare as good as empty at 10 p.m. We went into a charming little restaurant and chatted. Then, at home with fruits and cigars touchingly provided by Y[amamoto].

21st. Chrysanthemum festival at the imperial palace garden.[72] Great difficulty in procuring a fitting frock coat along with top hat. The former from unknown donor via Mr. Bärwald, who brought it personally;[73] the latter from Mr. Yamamoto; far too small, so I had to carry it around in my hand the whole afternoon. We were with the foreign diplomats,

die im Halbkreis angeordnet waren. Abgeholt
und begleitet von deutscher Botschaft.
Kaiserin schritt den Halbkreis von
innen ab, und sprach einige Worte mit
Männern und Weiben der Botschaften, mit
mir einige freundliche Worte französisch
Dann Erfrischungen im Garten unter Tischen,
wobei ich unendlich vielen Leuten vorgestellt
wurde. Garten wundervoll, künstliche Hügel,
Wasser, malerisches Herbstlaub. Chrysantemen
in Beeten wohlgeordnet wie Soldaten. Am
schönsten sind die hängenden Chrysanthemen.
Abends gemütlicher Abend bei Berliners
in reizendem japanischem Hause. Er
intelligenter Nationalökonom, sie grazröse,
intelligente Frau, ächte Berlinerin. Das Tanken
zen ist unter solchen Bedingungen
ermüdender als Arbeit, aber Inagaki's
halfen uns mit rührender Sorgfalt.
22. Um 10½ Uhr von Yukis und Y nach Kaigo
abgeholt zu Redaktionshäuschen. Angestellte
erwarteten uns festlich vor der Thür. Kam auch
schaftlicher Tan unvorkennbar Y. strahlte
mit seinen Kinderaugen unter grossem Hornbrille.

who were arranged in a semicircle. Picked up and accompanied by the German embassy. Jap[anese] empress stepped around the inside of the semicircle and spoke a few words with husbands and wives from the embassies, with me a few kind words in French.[74] Then refreshments in the garden at tables, where I was introduced to infinitely many people. Garden marvelous, artificial hills, water, picturesque autumn foliage. Chrysanthemums in booths properly lined up like soldiers. The hanging chrysanthemums are the most beautiful. In the evening, cozy evening at the Berliners' charming Japanese home. *He*, an intelligent political economist, *she*, a gracious, intelligent woman, true native of Berlin. Lazing about under such conditions is more tiring than working, but the Inagakis help us with touching solicitude.

22nd. Around 10:30, picked up by Gakis and Y[amamoto] for [meeting at] Kaizo's small publishing house.[75] Employees were waiting for us ceremoniously in front of the door. Comradely mood unmistakable. Y[amamoto] beamed with his childlike eyes under great big hornrimmed glasses.

Alle zusammen vor Redaktion in Gässchen
photographiert. Neugieriges Volk mit viel
Kindern schaute zu. Dann fuhren wir zu
prachtvollem Budd. Tempel. Blick in
Esssaal der Priester. Hundervolle Gebäude
mit herrlichen Schnitzereien. Priester sehr
freundlich, schenkte uns prächtiges Buch
mit Abbildungen von Kunstwerken. In den
Höfen übliche Photographieerei zum Teil
mit lustigen Schulmädchen (Zusammen aus Osaka),
die ebenfalls gerade den Tempel besuchten.
Dann Mittagessen im reizenden Hause Yamo-
moto. Ein prächtiger Mensch. Er beherbergt
in seinem Häuschen neben Frau und Kinderchen,
drei Dienstmädchen und einem Diener vier Studenten.
Wie friedlich und anspruchslos müssen diese
Leute sein! Nachmittags sahen wir Bauernhaus
und andere ganz einfache Japanerhäuschen,
alles blitzblank und freundlich. Viele gut
gehaltene lustige Kinderchen, abgehärtet gegen
die Kälte. Besuch beim Vorsitzenden der
Akademie. Ein Sohn desselben entpuppt
sich als Student des Züricher Polytechnikums
und Schüler von H. F. Weber. Verstimmung

12. Einstein and Elsa at residence of *Kaizo* publisher Sanehiko Yamamoto, Tokyo, 22 November 1922
(courtesy of the Leo Baeck Institute, New York).

Everyone from the publishing house photographed together in the alleyway.[76] Curious crowd of onlookers with many children. Then we drove to the magnificent Budd[hist] temple.[77] Glimpse into the monks' dining hall. Wonderful building with splendid carvings. Monks very friendly, gave us a magnificent book with illustrations of artwork. In the courtyards, usual photographing, partly together with cheerful schoolgirls from Osaka, who were also just visiting the temple. Then lunch at Yamamoto's charming house.[78] A splendid person. In addition to wife and young children,[79] he houses in his little home three maids and a servant, four students. How peaceful and undemanding these people must be! In the afternoon we saw a farmhouse and other very simple Japanese cottages, everything spotless and welcoming. Many well-cared-for, cheerful little children, inured to the cold. Visit to the chairman of the academy. One of his sons turns out to be a student at the Zurich Polytechnic and pupil of H. F. Weber.[80] Upset

wegen Empfang bei Fucisawa. Abends grosser
Empfang im deutsch-ostasiatischen Verein.
Gespräch mit vielen Deutschen und Japanern
nach dem Essen. Karussel im Kopf, aber
viel gelernt und enthusiastische Freundlich-
keit erfahren. Japanische Gelehrte viel
Sympathie für Deutschland. Japaner
haben mit Hilfe deutscher Ingenieure
selbständige optische Werkstätten errichtet.
23. Von 9 bis 11 Uhr Akten der von der Gesannt-
schaft verfolgten Frau des Angestellten
Schulz studiert. Armes Weib, das einer
Mandalverhüllung zum Opfer gebracht worden
soll. 11-10 ¾ 12-1 Uhr possierliches Gespräch
mit japanischen Journalisten über japanische
Eindrücke. Sitzen im Kreise Frag- und Antwort-
spiel. Dann gemeinsames lukullisches Essen
von 2-4 japanisches Konzert in der Musikschule
klagendes Flötenunisono mit viel Figürchen
ohne eigentliche Melodie. Harfe mit wenig Saiten
und mandolinenähnliches Instrument
mit Litanei-ähnlichem Gesang. Auch
ein Naturlaute nachahmendes Stück für Zupfinstrumente war darunter, graziös, melodisch arm,

regarding reception at Tujisawa's.[81] In the evening, grand reception at the German–East Asian Club.[82] Conversations with many Germans and Japanese after dinner. My head is spinning, but learned much and encountered enthusiastic friendliness. Japanese scholars have much sympathy for Germany. Japanese erected independent optical workshops with the help of German engineers.[83]

23rd. From 9 to 11 a.m., studied files on the persecution of the employee Schulz's wife by the embassy. Poor woman who is supposed to be sacrificed in a scandal coverup.[84] ⟨11–10:30⟩ 12–1 p.m. comical conversation with Japanese journalists about Japanese impressions.[85] Sitting in a circle, question-and-answer game. Then, sumptuous communal meal. From 2–4, Japanese concert at the music school, with lamenting flutes in unison, brief figures without proper melody.[86] Harp with few strings and mandolin-like instrument with litany-like singing. There was also a piece for plucked instruments mimicking the natural lute, graceful, melodically meager.

Von eigentlicher Gliederung und Harmonik keine Spur. Abends Essen bei Fujisava mit Nagaoka und ein paar anderen (auch einer von deutscher Botschaft darunter). Zierliche Töchterlein des Hausherrn. Gemütliche Gesellschaft mit harmlosem Gespräch. Vorher Besuche bei Berwald und Prof. Nagai (Chemiker).

24. Vormittags Spaziergang zu Fuss mit Inagaki. Essen im japanischen Wirtshaus. Bodensitzen schon üblich. Gebratene Hummern, schade um die Geschöpfe. Niedlicher Betrieb. Feine, stille Manieren des Publikums. Nachmittags Besuch in künstlerischem jap. Teehaus (Negi). Wundervolle altjapanische Bilder von feinster Rhythmik und Farbe. Schönes Tempel für japanische Psyche. Chinesisch-buddhistischer Einfluss Volksseele nicht entsprechend, mutet barock an gegen das originalen Erzeugnisse eigener Kunst des Landes. Vortrag 5½-7 und 8-10 zweiter allgemeiner Vortrag mit Ishiwara. Ungeheures Interesse des Publikums. 24-1.III Wissenschaft. Vorträge über Relativität u.d. Universität.

25. Vormittags Besuch von verrücktem Amerikaner, welcher andere Verrückte heilen zu können glaubt. 2 Uhr Besuch des physikalischen Instituts. Dann erste wissenschaftliche Vorlesung. 3½ Uhr

Not a trace of proper structure and harmonics. Evening meal at Fujisawa's with Nagaoka and a couple of others (someone from the German embassy among them as well). Dainty little daughter of the host. Pleasant company with harmless chitchat. Beforehand visits to see Bärwald and Prof. Nagai (chemist).[87]

24th. Morning stroll on foot with Inagaki. Meal at a Japanese inn.[88] Sitting on the floor difficult. Roasted lobsters; poor creatures. Charming establishment. Fine, quiet manners among the customers. Afternoon visit to artistic Japanese private home (Nezu).[89] Wonderful antique Japanese pictures of most refined rhythm and color. Beautiful evidence of the Japanese psyche. Chinese-Buddhist influence not in accord with the national soul; appears baroque compared to the country's own original artwork. From 5:30–7 and 8–10, second public lecture with Ishi[wara]. Immense interest by the audience.[90] 24th–1st Dec. scientif[ic] talks about relativity at the university.

25th. Visit by crazy American woman who believes she can cure other nuts. 2 p.m., visit at the physics institute. Then my first scientific lecture.[91] 3:30,

Begrüssung der Studenten in der Aula. Tiefer Eindruck,
ich sprach über Wissenschaft als internationales Gut.
6-8 Japanisches Theater. Wieder Chor mit Zasten-
und Schlag-Instrumenten. Ballet-artige Behandlung von bekannten Kindermärchen. Hoch interessantes Ge-
Berdenspiel - zum Teil sehr fremdartig. Abends
8½ - 10 Essen in japanischem Gasthaus, eingeladen
von Gesamtheit der Journalisten mit mehreren
Geisha. Diese führten zierliche Tänze mit Musik
auf. Tänzerinnen sehr jung; ältere Geisha mit
sehr ausdruckvoll sinnlichem Gesicht, unvergesslich.
Rede und Antwort in Brüsseln. Puppenessen für
Menschen. Dann wurden wir höflich verabschiedet,
damit der zwanglosere zweite Teil beginnen konnte.
Zwiegespräch mit dem kleinen Ina über Geisha, Moral etc.
26 (Sonntag) Okura Museum mit herrlichen chinesischen
und japanischen Statuen, Gemälden, Reliefs
und herrlicher hügeliger Gartenanlage. Nachmittags
Noh-Spiele. Antikes Drama mit japanischem Chor
Ganz langsame Bewegungen u. Masken. Hohe dramatische
Wirkung. Dann Besuch in riesigem Büchergeschäft
Alle 250 Angestellte vorhanden. Neugieriger Empfang
Essen auf Zimmer. Geschäftliches mit Hauns etc.
etc.

reception by the students in the auditorium. Profound impression; I spoke about science as an international asset.[92] 6–8, Japanese theater. Again chorus with string and percussion instruments. Ballet-style treatment of familiar children's fairy tales. Very interesting pantomime—in part very exotic.[93] In the evening from 8:30–10, dinner at a Japanese restaurant, invited by the journalists as a group, with a number of geishas. The latter performed charming dances with music. Dancers very young; older geishas with very expressive, sensual faces; unforgettable. Questions-and-answers in chest tone. Doll-size fare for humans. Then we were politely dismissed so that the more unconstrained second part could begin.[94] Private conversation with little Ina[95] on geishas, morals, etc.

26th (Sunday). Okura Museum with splendid Chinese and Japanese statues, paintings, reliefs, and gorgeous hilly garden.[96] Noh plays in the afternoon. Antique drama with Japanese chorus. Very slow movements and masks. Highly dramatic effect.[97] Then, visit to a gigantic bookstore. All 250 employees present.[98] Welcomed with curiosity. Meal in room. Business discussion with Yamamoto.

26th.

 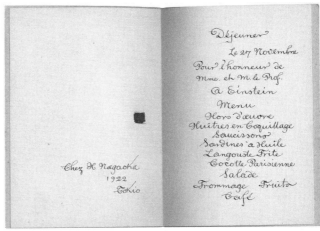

13. Menu for luncheon at physicist Hantaro Nagaoka's residence, Tokyo, 27 November 1922 (courtesy of Princeton University Library).

27. Besuch von Nagaokas Schwiegersohn mit reizendem Fräulein. Essen bei Nagaoka mit Gespräch über endlose japanische Univ. Examina. Dann Vorlesung über Tensortheorie. Dann Abendessen bei Tokugawa. 2 Musikstücke Gesang 2 Zupfinstrumente Tokug. Flöte. Inhalt: Landschaftseindruck. ... anmutige Tänze von Tok. Kinderchen. Dann meisterhafte Tanzpantomime einer Dame mit Begleitung von zwei singenden und zupfenden Frauen. (... letzter Durch mit Kinderchen) Abendessen prunkvoll, japanisch, gekocht von des Teufels Grossmutter. Dann Geigenspiel von mir (Gluck, Pfitzner, Bach) und japanischer Dame (Wsendowski)

28. Begrüssung in Handels-Hochschule. Prächtige Rede eines Studenten in Deutsch. Ich antwortete über Originalität der japanischen Kulturwerte. Tiefer Eindruck. Univ.-essen. Festlook, ohne Rede. Sass neben Rektor. Nachmittags Vorlesung und physikalisches Seminar (Vortrag über Karman'sches Problem) Kaibos-Verlag - Angestellte. Gemeinsames Abendessen in Bahnhofrestaurant & Rede (ich jüngster Angestellter des Verlags). Begrüssung durch feierlichen Händedruck des thatsächlich jüngsten Angestellten. Noch Artikelchen über japanische Musik diktiert.

14. Einstein lecturing at Tokyo University of Commerce, 28 November 1922 (courtesy of the Hitotsubashi University, Tokyo).

27th. Visit by Nagaoka's son-in-law with charming little wife. Meal at Nagaoka's with conversation about endless Japanese univ[ersity] examinations.[99] Then, lecture on tensor theory.[100] Then, supper at Tokugawa's.[101] Two pieces of music, vocal, two plucked instruments, Tokug[awa] flute. Content: landscape impressions. Pantomime dances by Tok[ugawa]'s little children. Then masterful pantomime dances by a woman, accompanied by two singing and plucking women, the latter also with little children. Supper ostentatious, Japanese, cooked by the devil's grandmother. Then, violin playing by me (Gluck, Hauser, Bach) and Japanese lady (Wieniawski).[102]

28th. Reception at University of Commerce.[103] Grand speech by one student in German. I replied with speech on the originality of Japanese cultural values. Deep impression. University meal. Festive, without speech. Sat next to university rector.[104] Afternoon lecture and physics seminar (talk about Kármán problem).[105] Kaizo Publishers employees. Joint dinner in train station restaurant. Speech (by me, as the publisher's youngest employee!). Welcomed ceremoniously by handshake with the actual youngest employee.[106] Also dictated a short article about Japanese music.[107]

29. Im Hand empfang Karte des Pfarrers Steineder *Schnell ansehen und Rückbilder halb in seiner Gegenwart* wegen Frau Schulze für Information. Dann der englische Arzt Gordon – Munroe, Arzt der Schulze. Konstatierte Psychose der Frau durch Misshandlung von Seite ihres Mannes (An-gestellten der deutschen Botschaft. 10½ Uhr Theezeremonie in feinem japanischen Hause. Genau vorgeschriebene Zeremonie, eine Mahlzeit zum Feier ~~~~~~~ der Freund-schaft. Bezeichnende Lebenskultur der Japaner. Der Herr hat 4 dicke Bände über die Zeremonie geschrieben, die er uns mit Stolz zeigte. Dann Empfang von 10000 Studenten der von Okuma (?) gegründeten, im demokratischen Geiste geleiteten Waseda – Universität mit Ansprachen. Mittagessen im Hotel. Dann Vorlesung. Besichtigung des Instituts. Interessan-te Mitteilung über Lichtbogen – Linienverschiebung. Um 6½ Uhr Begrüssung durch nätlayogische Gesellschaften. Beim Abschied aussen Begrüssung durch Seminaristinnen. Liebliches, heiteres Bild mit Gedränge im Halbdunkel. Zu viel Liebe und Verehrung für einen Sterblichen. An-kunft zuhause todmüde.

15. Einstein and President Masasada Shiozawa at Waseda University, Tokyo, 29 November 1922 (cour-tesy of Waseda University).

29th. While in shirtsleeves received a card from pastor Steinichen announcing his visit to keep me informed about Frau Schulze's case. Changed and got dressed quickly, partly in his presence. Then the English physician Gordon-Munroe, who was attending Frau Schulze.[108] Verified that wife's psychosis is due to husband's maltreatment (employee at the German Embassy). 10:30, tea ceremony in a fine Japanese home. Exactly prescribed ceremony for a meal to celebrate friendship. Glimpse into the contemplative cultural life of the Japanese. The man has written four thick volumes on the ceremony, which he proudly showed us.[109] Then, reception by 10,000 students at Waseda University, founded by Okuma (?) in the spirit of democracy, with addresses.[110] Lunch at hotel. Then lecture.[111] Viewing of the institute. Interesting communication about electric-arc line-shift. At 6:30 reception by pedagogical societies. During farewells, greeting by female seminar participants outside. Sweet, cheerful scene of throngs in semi-darkness.[112] Too much love and spoiling for one mortal. Arrival home dead tired.

zur Reise - Information

30. Mit Frau aus Bahnhof. Einziges Mal alleine ausgegangen. Komische Schwierigkeit der Verständigung. Inayaki sucht und findet uns unterwegs per Auto und führt mich zu kaiserlicher Kapelle, um altchinesische Musik zu hören (10½ - 1½), die nur dort noch durch Tradition am Leben erhalten wird. Gemeinsame indische Wurzel der byzantinischen und chinesisch-japanischen Musik. Choralähnliche wunderbare Klangmalerei. Flöte, Zupfinstrumente, Zungeninstrumente, auch für sehr hohe Töne, die silbrige Klangwirkung geben. Vortrag in Univ. Diskussion. Tanaka Aufklärung von Inkonsequenz von Dai's theoretischem Versuch. Feierliche Begrüssung von Abordnungen der Studenten (im Ganzen ca 20000) der Universitäten von Tokyo. Abendessen bei der Botschaft. Diplomaten und sonstige grosse Tiere. Herrliche Musik. Aber sonst fad und steif. Ich stümperte auch ein wenig Geige, sehr schlecht aus Müdigkeit und Mangel an Übung. Der Zwischen Botschafter brachte uns heim; gemütliches Ehepaar.

1. Dezember. Essen mit Ehepaar Witz, das gestern am Bahnhof getroffen. Mitteilungen über Gefangenschaft in Kanada. Letzter Vortrag über kosmologisches Problem. Dank der zugelassenen Studenten. Riesiges Abendessen

16. Einstein lecturing on relativity using variational calculus at the Department of Physics, Tokyo Imperial University, early December 1922 (with permission of the Estate of Kenji Sugimoto and Kodansha Ltd. and courtesy of the Albert Einstein Archives).

30th. With wife to the tourist information at train station. Sole excursion on our own. Funny difficulties in communicating. Inagaki seeks and finds us along the way by car and drives me to imperial ensemble to listen to ancient Chinese music (10:30–1:30), which is only being kept alive there by means of tradition.[113] Common Indian roots for Byzantine and Sino-Japanese music. Chorale-like. Wonderful onomatopoeia. Flute, plucked instruments, reed instruments, also for very high pitches, producing a silvery tone. Lecture at univ[ersity] Discussion: Tamaru elucidation of inconsistency of Doi's theoretical experiment. Festive address by student delegations (in total, approximately 20,000) of the universities of Tokyo.[114] Dinner at the embassy. Diplomats and other big shots. Gorgeous music. But otherwise boring and stuffy. I bungled through some violin as well; very badly due to fatigue and lack of practice. Danish ambassador brought us home; good-natured married couple.[115]

1st December. Meal with Mr. and Mrs. Witt, whom we met yesterday at the train station. Information about imprisonment in Canada. Last lecture on cosmological problem. Thanks from admitted students.[116] Huge dinner

im Hotel. Ganze geistige Elite anwesend.
Nach dem Essen musste ich (nach Yam_____)
Ansprache halten und – geigen (Kreutzersonate).
(Vormittags Chemiker Tamaru Besuch im Hotel).
2. Besuch der technischen Schule. Begrüssung der
Studenten Ansprache Takeuchi. Fahrt nach
Sendai (1–9). Londu *Aichi* 4 Stunden entgegengefahren.
Ankunft. Kollegen, Rektor am Bahnhof, auch
Botaniker Molisch. Lebensgefährliches Gedränge
auf Weg zum gegenüberliegenden Hotel. Dort Empfang
der Behörden. Morgens Vortrag 9½ bis 12 u. 1–2½.
Fahrt mit Yam., dem Kurikat. Maler Okamoto
nach Kaefuminal. Wundervolle Küstenlandschaft.
Einkehr in japanischem Gasthaus auf japanische
Art. Abendessen mit Physikern im Hotel. Be-
kanntschaft des Dichters Tsuchii. Schenkte
mir Skizzenbuch von Hokusai und selbst-
geschriebenes italienisches Gedichtbuch. Abends
rührender Empfang in Universität. Studentenversammlung. Dann mit Professoren. Jass neben
Molisch und Dekan der med. Fakultät. *Musste Namen und Datum* *schrieb Tsuchii an eine* *Wand malen.*
4) Rückreise nach ~~Sendai~~ *Nicow* mit Dr. Yam. und Okamoto.
Hinda kam 1 Stunde mit. Prächtige Menschen.
Hitler, bescheiden, natur- u. kunstliebend.
Unvergesslich. Wundervolle Gebirgslandschaften

17. Einstein and Elsa at the Tokyo School of Technology, 2 December 1922 (with permission of the Estate of Kenji Sugimoto and Kodansha Ltd. and courtesy of the Albert Einstein Archives).

at hotel. Entire intellectual elite in attendance. After the meal I had to deliver an address (after Yamamoto) and—play violin (Kreuzer sonata). (In the morning, visit by chemist Tamaru at hotel).[117]

2nd. Visit to the School of Technology. Student reception. Address Takiuchi.[118] Travel to Sendai (1–9). Honda & Aichi travel four hours to join us.[119] Arrival. Colleagues, rector at train station, botanist Molisch as well. Life-threatening jostle on the way to hotel on opposite side of the street. There reception by authorities.[120]

3rd. In the morning, lecture 9:30 until 12 & 1–2:30.[121] Trip with Yam[amoto], the caricat[urist] painter Okamoto to pine island.[122] Spectacular coastal landscape. Stop at Japanese restaurant, Japanese-style. Dinner with physicists at the hotel. Make acquaintance with the poet Tsuchii. Gave me sketchbook by Hokusai and a book of Italian poems he had written himself.[123] In the evening, moving reception at the university. Student assembly. Then with professors. Sat beside Molisch and dean of the Med[ical] faculty. Had to write name and date in ink on wall.[124]

4th. Departure for ⟨Sendai⟩ Nikko with In[agaki], Yam[amoto], und Okamoto. Honda came along for one hour.[125] Superb people. Humorous, modest lovers of nature and art. Unforgettable. Splendid mountainous landscapes

vom Zug aus. Gestern und heute überall besondere Freundlichkeiten der Eisenbahnbeamten.
Frauen unterwegs verfehlt, weil sie der
Tokyo Zug versäumt. Malerische Fahrt. Gespräch
mit halb amerikanisierten deutschamerikanischen
Seidenstrumpf-Fabrikanten. Durch das Dorf
Nieco mit Ineg. und Okamoto zu Fuss ins Hotel.
Letzterer machte noch am selben Abend
mehrere sehr reizvolle Charakterskizzen.
5. Inag. Schwer aus dem Bett zu kriegen, weil
wieder bei einer Frau. Gegen 9½ Aufbruch
nach dem 1300 m hoch gelegenen Tempel-See.
Auto bis zum eigentlichen Aufstieg, dann
Aufstieg durch prächtige Wälder mit herrlichen
Ausblicken, auf Gebirge, enge Thäler und Ebene,
Oben heftiger Schneesturm, bei empfindlicher
Kälte, der uns bis unten treu blieb. Okamoto,
der ärmste in Strohsandalen, aber stets humorvoll und schalkhaft. Abends x des Telegramm
von deutscher Gesellschaft in Kobe. Zum
mindesten in Japan hab ichs lieber mit
Japanern zu thun. Sind ähnlich den Holzenern
in ihrem Temperament, aber noch feiner, noch
ganz durchdrungen von ihrer künstlerischen Tra-

seen from train. Yesterday and today everywhere special favors by railway officials.
Failed to meet wives along the way because
they missed the train in Tokyo.[126] Picturesque
trip. Conversation with semi-Americanized
German-American silk-stocking manufac-
turer. Through the village of Nikko with Inag[aki] and Okamoto on foot to the hotel.
The latter sketched a number of very charming caricatures that same evening.[127]

5th. Hard to get Inag[aki] out of bed because reunited with his wife.[128] Around 9:30
set out for the temple lake at 1,300-m alti-
tude.[129] Car ride up to actual ascent. Then
hike through magnificent forests with splendid views onto mountains, narrow valleys,
and plateaus. At top, hefty snow storm with
severe coldness that faithfully accompanied

18. Sketch by Ippei Okamoto, "Albert Einstein
or The Nose as a Reservoir for Thoughts,"
4 December 1922 (courtesy of the Estate of Ippei
Okamoto).

us down to the bottom. Okamoto, poor blighter, in straw sandals, but
always full of humor and mischief. In the evening, umpteenth telegram from the German Society in Kobe. At least in Japan, I much
prefer dealing with Japanese. They are similar to Italians in temperament, but even more refined, still entirely soaked in their artistic

...stran, nicht nervös, humorvoll. Unterwegs Gespräche über buddhistische Religion. Gebildete Japaner liebäugeln mit Urchristentum. Ferner Gespräche über japanische Weltansicht vor Anschluss an Europa. Es scheint, dass sich die Japaner keine Gedanken darüber machten, warum es auf ihren Südinseln heisser sei als auf ihren Nordinseln. Auch scheint ihnen nicht zum Bewusstsein gekommen zu sein, dass die Sonnenhöhe von der Nordsüdlage abhängig ist. Intellektuelle Bedürfnisse scheinen bei diesem Volk schwächer gewesen zu sein als künstlerische. Naturanlage?

6. Tempelsystem von Nikko besucht Natur und Architektur prachtvoll vereint. Zeremonielle Steigerung durch System von Höfen. Zentrale Gebäude wunderbar durch farbige Schnitzereien verziert. Etwas überladen. Freude an Darstellung von Natur überwiegt das Architektonische und erst recht das Religiöse. Lange Vortrag von Triester über Historisches – nicht zum Erleben. Wundervoller Steintreppen-Aufstieg unter Cedern zum Grab des ältesten Tokugawa. Nachmittags Bruder Beck mit Tochter da. Fussweg nach Bahnhof durch diesem Gebirge bei sinkender Sonne in prächtiger

tradition, not nervous, full of humor. Along the way, conversations about Buddhist religion. Educated Japanese flirt with primitive Christianity. Additionally, conversation about Japanese worldview prior to contact with Europe. It seems that the Japanese never thought about why it is hotter on their southern islands than on their northern islands. Nor do they seem to have become aware that the height of the sun is dependent on the north-south position. Intellectual needs of this nation seem to be weaker than their artistic ones—natural disposition?[130]

6th. Visited temple system at Nikko. Nature and architecture magnificently united. Cedar avenue. Enhancement through system of courtyards. Central buildings marvelously decorated with colorful carvings. Somewhat overdone.[131] Joy in representing nature outweighs the architectonic and, even more so, the religious. Long talk by priest about historical matters—deadly. Wonderful stone stairway under cedars leading to the grave of the eldest Tokugawa.[132] Afternoon, Brother Beck with daughter here. On foot to train station with these mountains fabulously illuminated by the setting sun.

Beleuchtung. Fahrt nach Tokyo. Dort grosse Hetzerei mit Packen im Hotel.

7. Noch grössere Hetz mit Krach, Kofferschliessen. Bärwald auch da. Fahrt nach Bahnhof. Entgeltung Abschied von Tokyo. Fahrt mit Ishiv., Inayakis, Yamamoto & Frau nach Nagoya. Ich mit Artikel-schreiben beschäftigt, sehr pressant, über japanische Eindrücke. Ankunft begrüsst von grosser Schar Studenten und Schüler. Gemütliches Abendessen der ganzen Gesellschaft mit vier Leuten von Koiyoschen-Verlag im Wirtzimmer, mit Ahorn geschmückt. Hock in Inayakis Zimmer. Michaelis im Hotel getroffen.

8. Morgenspaziergang in Hauptstrasse Nagoya bis Bahnhof, vergeblicher Versuch, Pfeifentabak zu kaufen. Mit Yam, Inay & Ishiwara Besuch von schintoistischem Tempel. Grosser Hain. Darin Tempelanlagen nach Hofsystem. Elegante Schmuckloser hier Ausser ... Decken. glatte Holzbauten. Muss von Süden gekommen sein. Charakteristischer Dachansatz

Trip to Tokyo.[133] There, great scurry while packing in the hotel.

7th. Even greater rush with quarrel, suitcase shutting. Bärwald also there. Trip to train station. Final departure from Tokyo. Travel with Ishiw[ara], Inagakis, Yamamoto & wife to Nagoya. I'm busy with article writing, very hurriedly, about Japanese impressions.[134] Arrival, met by huge crowd of students and pupils. Cozy supper with the entire company with four people from Kaizosha's publishing house in tavern room decorated with maple.[135] Get-together in Inagakis' room. Met Michaelis in the hotel.[136]

⟨8th⟩ 9th. Morning walk along the main street of Nagoya up to train station; futile attempt to buy pipe tobacco. With Yam[amoto], Inag[aki], & Ishiwara, visit Shinto temple.

19. Einstein on his arrival at Nagoya railway station, 7 December 1922 (with permission of the Estate of Kenji Sugimoto and Kodansha Ltd.).

Large grove. In it, temple complex with courtyard system. Elegant, smooth wooden structures. Unadorned. Empty booths for souls.[137] Must have come from the south. Characteristic roof structure.

Naturreligion, vom Staate benutzt. Viel
Götterei. Ahnen u Kaiserkultus. Bäume
Hauptsache bei Tempelanlage. Fahrt nach
Kioto nach grossem Bahnhofabschied der
Studenten und Lehrer. Kioto freundlicher
Empfang der Physiker und Studenten.
8) Besuch des kaiserlichen Schlosses mit
   herrlichem Festungsgebäude (turmartig)
   Im Schloss prächtige Naturmalereien an
   den Wänden und Thüren. Figur, Winterzimmer,
   Pflanzen und Vögel. Hofscenen. Nachmittags
   Musizieren mit Michaelis, grosser Vortrag in
   Zirkus mit Ishawara.
11. Fahrt nach Osaka (grosse Fabrik- und Handelsstadt)
    Empfang von Bürgermeister u Studenten am
    Bahnhof. Im Hotel grosse Honoratioren
    Vorstellung mit gehaltvollen Händedrucken.
    Festessen von Prof. Sata in grossem Saal
    Militär Trompetenmusik, riesiges Essen.
    Reden mit viel Pathos, auch von mir.
    (Japanisch gemildertes Amerika). Von
    5½ - 7 und 8 bis 9 ½ Vortrag. Das Ganze nicht
    so furchtbar anstrengend, weil alle Leute
    rücksichtsvoll und bescheiden. Bei Hämbeln
    grosse Entrüstung der zu Hause gelassenen Gattin.

Natural religion, utilized by the state. Polytheism. Worship of ancestors & emperor. Trees the principal thing in the temple complex. Trip to Kyoto following grand farewells at station by students and teachers. Kyoto, friendly welcome by univ[ersity] physicists and students.[138]

8th.[139] Visit of the imperial palace with magnificent fortified building (towerlike). Inside the palace, splendid paintings of nature on the walls and doors. Tigers, winter room, plants and birds. Courtly scenes.[140] Afternoon, music making with Michaelis, major lecture at the circus with Ishiwara.[141]

⟨10⟩ 11th.[142] Trip to Osaka (large factory and commercial town). Met by mayor & students at train station. At the hotel, 11:30, important dignitaries. Introduction with substantial handshaking. Banquet [hosted] by Prof. Sata in great hall. Military trumpet music, immense meal. Speeches with much pathos, also by me.[143] (mitigated America à la Japan.) From 5:30–7 and 8 until 9:30, lecture.[144] The whole thing not so terribly strenuous because everyone considerate and modest. Upon return home, great indignation by left-at-home wife.

10. 10½ – 12 und 1–3 Vortrag in Kioto. Grosse
Kälte in herrlichem Saal. Dann Besuch von
kaiserlichem Garten und Krönungsschloss.
Der Schlosshof gehört zum Schönsten,
was ich je an Architektur gesehen.
Kaiser Allüren eines Gottes, für ihn sehr
unbequem. Im Saal, wo vom Hof aus
die Krönungssessel sichtbar sind, ¹ Bildnisse
von etwa 40 — chinesischen — Staatsmännern
als Anerkennung für die kulturelle
Befruchtung, welche Japan durch
China empfangen hat. Diese Verehrung
für fremde Lehrer lebt heute noch
unter den Japanern. Rührende Anerkennung vieler Japaner, die in Deutschland studiert haben, an ihre deutschen
Lehrer. Es soll sogar zur Erinnerung
an den Bakteriologen Koch ein Tempel
existieren. Ernste Hochschätzung ohne
eine Spur Zynik oder auch nur Skepsis
für Japaner charakteristisch. Reine Seelen
wie sonst nirgends unter Menschen —
Man muss dies Land lieben und verehren.

¹ Gang von Bauten. Umgebung. Austrags-
raum und Kehrseite offen gegen
sandbedeckten Hof.

10th. 10:30–12 and 1–3, lecture in Kyoto. Very cold in splendid hall.[145] Then, visit of imperial garden and coronation palace.[146] The inner palace courtyard is among the most exquisite architecture I have ever seen. Completely surrounded by buildings. Audience chamber and coronation hall open out onto sand-covered courtyard. Emperor [has] status of a god; for him very uncomfortable. Inside the hall, where the coronation thrones are visible from the courtyard, portraits of about 40—Chinese—statesmen in acknowledgment of the cultural influence Japan has received from China.[147] This veneration of foreign teachers still lives on today among the Japanese. Touching recognition among many Japanese who have studied in Germany for their German teachers. There is supposedly even a temple in memory of the bacteriologist Koch.[148] Ernest respect without a trace of cynicism or even skepticism is characteristic of Japanese. Pure souls as nowhere else among people. One has to love and admire this country.

12. Vormittags Kirschnaufen. 2 Uhr altes Tokugawa-Schloss mit wunderbaren Landschaftsbildern (Wolken, Bäume, Vögel, drollige Tiger, auf Goldgrund gemalt) Gemälde von Balken unterbrochen, sodass Wände weggetäuscht und märchenhaft farbenschöne äussere Fortsetzung des Innenraumes vorgetäuscht wird. Mit Ishiwara Rechnung über Energietensor des elektromagnetischen Feldes in isotropen ponderabeln Substanzen für gemeinsame Abhandlung in japanischen Akademie-Berichten.

13. Reise nach Kobe. Mittagessen mit Okamoto Yamamoto und dem bedeutenden jungen Sozialpolitiker in Fischerdorf vor Kobe (Ausflugsort). Vortrag mit Ishiwara von 5 ½ bis 8 Uhr, Essen beim Konsul Trautmann. Dann Empfang im Deutschen Klub. Heimfahrt mit Frau allein im Bummelzug. (Ankunft Kioto 4 Uhr Nachts)

14. Festliches Mittagessen mit Professoren der Universität. Grosse Studentenversammlung. Ansprache des Rektors und Vertreters der Studentenschaft in tadellosem Deutsch sehr herzlich. Dann Vortrag von mir über Entstehung der Relativitätstheorie (auf Wunsch). Besuch im physikal. Institut. Höchst interessant, besonders Kimuras Untersuchungen über Verbreiterung der Spektrallinien.

12th. In the morning, a breather. 2 p.m., old Tokugawa palace with beautiful landscape paintings (clouds, trees, birds, droll tigers, painted on gold background). Painting interrupted by beams, makes walls seem to vanish and the interior to extend into fabulously colorful outdoors.[149] With Ishiwara, calculation on energy tensor of the electromagnetic field in isotropic ponderable substances for joint article in Japanese Academy proceedings.[150]

13th. Trip to Kobe. Lunch with ⟨Okamot⟩ Yamamoto and the important young labor politician[151]         in fishing village near Kobe (tourist attraction). Lecture with Ishiwara from 5:30 until 8 p.m., dinner at consul Trautmann's. Then reception at the German Club.[152] Trip home alone with wife on local train (arrival Kyoto, 1 a.m.).

14th. Festive lunch with professors from the university. Large assembly of students. Address by rector and representative of the student association in impeccable German (very cordial).[153] Then lecture by me on the genesis of the theory of relativity (by request).[154] Visit to the phys[ics] institute. (Highly interesting, particularly Kimura's investigations on broadening of spectral lines.[155]

Abends kommt Nagaoka von Tokyo mit Koffer voll prächtiger Geschenke von der Universität Tokyo.

15. Abschied von Nagaoka. Besuch von herrlichen buddhistischen Tempel Gedächtnis-Gottesdienst für Tote. Freundlicher Empfang durch Priester. Besichtigung der grossen Glocke; Klöpel horizontal und aussen. Blühender Kirschbaum vor dem Tempel. Photographieren im phys. Unov. Institut mit kleinem Empfang und Kolloquium. Dämmerung Besuch eines auf hohen Pfählen am Bergabhang stehenden schintoistischen Tempels. Dann Besuch festlich beleuchteter Strasse mit Luxusgeschäften und grossem Betrieb. Unbeschreiblich heiteres Bild wie Oktoberfest. Gewimmel von Laternchen und Fähnchen. Strasse ungeheuer dreckig, alles andere blitzblank und farbenprächtig.

16. Besuch buddhist. Tempels am Fuss des Hügels. Wundervolle Architektur subtrop. üppige Vegetation. Westtempel mit prachtvollen Gemälden. Auch harmonische Behandlung menschlicher Figuren in der Landschaft, an ital Renaissance erinnernd. Niemals Portraits oder Gruppenkomposition. Nachmittags Biva-See mit wundervoll gelegenen

In the evening, Nagaoka arrives from Tokyo with a suitcase full of magnificent presents from the University of Tokyo.[156]

15th. Take leave of Nagaoka. Visit to splendid Buddhist temple, memorial mass for the dead. Friendly welcome by monks. Tour of the big bell; clapper horizontal and on the outside. Blossoming cherry tree in front of the temple.[157] Photographs taken at the univ[ersity] phys[ics] institute with small reception and colloquium. At dusk, visit to a Shinto temple positioned on high stilts on the side of a mountain. Then visit of festively lit street with luxury shops and much goings-on. Indescribably cheerful scene, like Oktoberfest. Swarm of lanterns and little flags. Street surface extremely dirty, everything else sparkling clean and brightly colored.[158]

20. Einstein on the steps of the Chion-in Temple, Kyoto, 15 December 1922 (with permission of the Estate of Kenji Sugimoto and Kodansha Ltd.).

16th. Early visit to [a] Buddhist temple at the foot of the ⟨mountain⟩ hill near to the hotel. Wonderful architecture, subtrop[ical] lush vegetation.[159] In the morn[ing], western temple with magnificent paintings. Also harmonious treatment of human figures in the landscape, reminiscent of Ital[ian] Renaissance. Never portraits or group compositions.[160] In the afternoon, Lake Biwa with wonderfully situated

und architekt. sehr vollkommenen alten Felstempel.
Abends mehrere/viele Briefe geschrieben

17. Besuch mit Frau im Seidengeschäft. Herrliche
Landschafts- & Tierstickereien. Nachmittags
allein auf Hügel gestiegen zu Sonnenuntergang
Japanischer Wald (ohne) und Lichteffekte unvergleich-
lich. Abends nach Nara gefahren. Mit Yoki zu Fuss ins Hotel

18. Rundgang im Tempelbezirk. Zahme Rehe
wimmeln herum und schnubbern an einem
herum in der ganzen Landschaft. Tempel
architektonisch prachtvoll. Tempel mit
grosser Buddafigur besonders majestätisch
(mehr als 1000 Jahre alt), die Figur ziemlich
roh. Nachmittags Staatsmuseum von
alten Skulpturen. Um bei rührend schöne
Dinge aus der Zeit 700 – 1200. Tiefer Eindruck
von Japanischer Charakterisierungskunst.

19. Mit Inayaki auf kahlen Hügel (des jungen
Grases) gestiegen, der im Japaner viel Seligkeit
auslöst, indem er das Frühlingswerden verkörpert
Nachmittags Briefe & Karten, auch Reale-
nung für (Oshisawa). Von 6 Uhr Ab. bis 6 Uhr Morgens
Reise nach Miyajima.

20. Ankunft bei finstrer Nacht. Gebadet, ins Bett und

and architect[urally] most perfect ancient rock temple.[161] In the evening, wrote several letters.

⟨19th/18th⟩ 17th. Visit silk store with wife. Splendid landscape & animal embroidery.[162] In the afternoon, climbed hill alone at sunset. Japanese forest (maple) and light effects unrivaled. In the evening, traveled to Nara. With Gaki on foot to hotel. Very tasteful, semi-Japanese style and excellent.[163]

⟨19th⟩ 18th. Tour around the temple district. Tame deer milling about and nuzzling one throughout the landscape. Temple architecturally magnificent. Temple with large Buddha figure especially majestic (more than 1,000 years old); the figure quite unrefined.[164] Much superstition. Lanterns, memorial stones. Bits of paper on trees and by temples. In the afternoon, state museum of ancient sculptures.[165] Some charmingly pretty objects from the period 700–1200. Profound impression of Japanese art of representation.

⟨20th⟩ 19th. With Inagaki climbed up bald hill (of fledging grass) that delights the Japanese as it embodies the bliss of spring.[166] ⟨In the evening⟩ In the afternoon, letters & postcards, also calculation for joint paper with Ishiwara. From 6 p.m. until 6 a.m., trip to Miyajima.[167]

⟨21th⟩ 20th. Arrival in the dark of night. Bathed, went to bed and

bis 10 geschlafen. Von 11–12 bezaubernder Küstenspaziergang
zu dem ins Wasser (Flutgebiet) gebauten Tempel mit
graziöser Pagode. Nachmittags mit Yaki Tour auf
den Gipfel des die Insel hauptsächlich bildenden
Berges. Wunderbarer Blick über japanische Binnensee.
Zarteste Farben. Unterwegs unzählige Tempelchen,
Naturgottheiten gewidmet. Steinfiguren oft entzückend.
Der ganze Weg aus Stufen aus gefrorenen Granitsteinen (etwa 20 cm Höhe)
Denkmal japanischer Naturliebe und allerlei liebenswürdigen Aberglaubens. — Mittags Depesche
von Solf wegen Dementierung von Hardens Behaup-
tung, ich hätte mich nach Japan in Sicherheit
bringen müssen. Meine Antwort: Angelegenheit zu
kompliziert für Telegramm, Brief folgt. Letzteren
schrieb ich abends, wahrheitsgetreu.

21. Küstenspaziergang bei strahlender Sonne. Telegramm
an Gemeinde Changhai. Okamoto dabei. Nachmittags Wald-
und Küsten-Spaziergang. Quallenjagd mit Steinen
Yaki und Okamoto.

22. Yamamoto gekommen kleine Spaziergänge.
Holzklötzchenrätsel. Mit grosser Schwierigkeit
gelöst. Kleine Vergiftung mit offenem Kohlenfeuer (Steinkohlen darunter!)
im Zimmer, besonders das Frauen + ächtz
hergenommen.

slept until 10. From 11–12, enthralling walk along the coast to the temple with graceful pagoda, built in the water (tidewater region).[168] In the afternoon, tour to the peak of the mountain lending the island its main form, with Gaki. Wonderful view over the Japanese Inland Sea.[169] Subtlest of colors. Along the way, countless small temples, dedicated to natural deities. Stone figures often delightful. The entire path of steps hewn into granite rocks (height around 700 m). Memorial to Japanese love of nature and all sorts of endearing superstitions.—Noon, telegram from Solf about denial of Harden's assertion that I had to escape to safety in Japan. My answer: affair too complicated for telegram; letter to follow. I wrote the latter that evening, in accordance with the truth.[170]

21st. Coastal walk in brilliant sunshine. Telegram to the Community of Shanghai. Accompanied by Okamoto. Afternoon, walk in woods and along coast. Jellyfish hunt with rocks (Gaki and Okamoto).

22nd. Yamamoto arrived, brief walks. Wooden block puzzle. Solved it with great difficulty. Minor intoxication from open coal fire (partly anthracite!) in room; the women were particularly affected.

23. Fahrt nach Moji. Üppiger Empfang. Hatte bei Ankunft in Schimonoseki Examen mit Journalisten zu bestehen. Abends im Mizzel-Klub fürstlich untergebracht ausserhalb Moji.

24. Photographiert zum 10 000. Mal. Dann Fahrt nach Fukuoka unmittelbar vor Vortrag, der von 1–3 und 4–6 dauerte, der erkältete Ishiwara muss übersetzen. Studentenempfang abgesagt, weil die Leute sonst wie in Sendai auf Gratisvortrag rechnen und der arme Yamamoto lackiert ist. Nach Vortrag Kaishiosha-Abendessen, das nahe zu ewig dauert. Grossen Teil der Gesellschaft und einer nebenan plauderten von Gymnasial-Mathematik-lehrern durch Reiswein tüchtig beschwipst und sehr lustig. Prof. Myabe rührend, begleitet mich überall hin, zum Schluss in japanisches Hotel, wo die Wirtin eine rührende Freude hat und sich rad 10 mal kniend mit Kopf nach Boden verbeugt. Räume höchst geschmackvoll Wohnzimmer und 2 mit europäischen Sitzgelegenheiten provisorisch ausgestatteten Nebenzimmern, alles durch Papier-Schiebethürchen getrennt, so man bequem mit

23rd. Trip to Moji. Sumptuous reception. Had to pass examination with journalists upon arrival in Shimonoseki. In the evening, princely accommodations at Mitsui Club outside of Moji.[171]

24th. Photographed for the 10,000th time. Then, trip to Fukuoka immediately prior to lecture, which lasted from 1–3 and 4–6; Ishiwara, suffering from cold, had to translate. Student reception canceled because, otherwise, the people, as in Sendai, generally count on lecture being free of charge, and poor Yamamoto would be ripped off.[172] After lecture, Kaishosha dinner that almost lasts forever. A large part of the company and another group of high-school maths teachers at the next table, properly tipsy from rice wine and very funny.[173] Prof. Miyake thoughtfully accompanies me everywhere, in the end to a Japanese hotel, where the hostess of the inn is deeply thrilled and, on her knees, bows her head to the ground around 100 times. Rooms in exquisitely good taste. Living room and two adjoining rooms provisionally outfitted with European seating, everything separated by paper sliding doors that

kleinem Finger bethätigen kann.

25. Wilder Tag! Um 9 Uhr erscheint Kuwaki, Inagaki auch die Dame Wirtin, die drollige. Sie kommt mit etwa 6 seidenen Stoffstücken von etwa 3/4 M. Länge und einem Bündel Pinseln und japanische Tusche, und ich muss alles mit meinem Namen vollmalen. Gespräch mit Kuwaki über Molekulartheorie. Fragen der Relativität. Dann 11 Uhr Frauen von der Bahn geholt (Else & Frau Inagaki) fahrt ins japanische Hotel. Besichtigen der Stadt und vieler Läden. Um 1 Uhr zum Universitätsessen in Räumen der medizinischen Fakultät. Händeschütteln mit sehr vielen Professoren. Besichtigen von Gallensteinen, mikroskopischen Präparaten über Weil'sche Krankheit, Fischen, die das Resultat von Kreuzungen darstellen, alles eigens im Empfangssaal aufgestellt. Nach Mittagessen Rede von Präsident und mir. Viele Geschenke erhalten. Besuch eines Tempels, des physikalischen, festigkeits- mineralogischen Instituts, dann Besuch bei Myake, der viele Merkblätter Kinder hat und endlich in Gewerbeausstellung im Stadthaus, wo der immerinde Provinzgouverneur

a little finger could easily shift aside.[174]

25th. Wild day! Around 9 a.m. Kuwaki shows up, Inagaki also there. Then the funny innkeeper. She arrives with about six pieces of silk about ¾ m in length and a bundle of brushes and Japanese ink, and I am supposed to cover it all in paint with my name.[175] Discussion with Kuwaki about epistemol[ogical] questions on relativity. Then, 11 a.m. picked up women (Elsa & Mrs. Inagaki) from the train station. Trip to Japanese hotel.[176] Tour of the city and many shops. At 1 p.m. to university luncheon in rooms of the medical faculty. Shaking of hands with very many professors. Viewing of gall stones, microscopic preparations on Weil's disease, fish representing the result of crossbreeding, all specially set up in reception hall. After lunch, speeches by president and by me. Received many gifts.[177] Visit to a temple, the physics and solid mineralogy institute.[178] Then visit at Miyake's, who has four of the dearest children, and finally to a trade exhibition at the town hall, where the provincial governor in attendance

...gens wundervolle Gemälde hat aufhängen lassen. Dann Abreise, zu welcher alle kamen, auch der freundlichste aller japanischen Werkmänner. Ich aber war tot, und mein Teichmann fuhr nach Moji zurück, wo er noch in einer Kinder-Weihnacht geschleppt wurde und den Kindern vorgeigen musste. Endlich 10 Uhr Heimkehr, noch einige Briefe aus der Heimat, Bett.

26. Vormittags und abends Besteigung eines Hügels mit herrlicher Aussicht auf Berge und Meer. Yamamoto angekommen. Unannehmlichkeit, weil er uns noch nachlaufen muss, indem der Mizzi-Klub die Krallen zeigt und horrend verlangt. Vorrede zu Vorlesung geschrieben. Gedanke über $\left(\frac{\partial}{\partial x}\right)\left(R^a_{\ b,\ c\ m}\right)$

27. Spazierfahrt mit Mizzi-Dampfer auf chinesisches Meer. Spaziergang durch Schimonoseki. Abendessen mit Yamamoto & Watanabe. Nachher Nagai gekommen. Gedicht mit Zeichnung für Frau Yamamoto und viel Seide mit Tusche vollgeschmiert. Animierteste Stimmung. [Eitel Friede!]

28. Regnerischer Tag. Abends Einladung des kommerziellen Klubs von Moji. Ich zeige die Japaner

*[margin note]* Abends kommt Japaner mit Stoss Papier und will, dass ich ihn meine ... und zufälliges ... besehe!

21. The Einsteins at Christmas ceremony for children at the Moji YMCA, 25 December 1922 (with permission of the Estate of Kenji Sugimoto and Kodansha Ltd.).

had specially arranged to have wonderful paintings displayed.[179] Then departure, to which *everyone* came, including the friendliest of all innkeepers. But I was dead, and my corpse rode back to Moji where it was dragged to a children's Christmas and had to play violin for the children. At last, 10 p.m. return, supper, many letters from home, bed.[180]

26th. In the morning and evening, climb up a hill with beautiful view of mountains and sea. Yamamoto arrived. Embarrassed, because he has to relocate us, as the Mitsui Club is baring its claws and demanding horrendous sums. Wrote introduction to lecture.[181] Idea about $\left(\frac{\partial}{\partial x_\alpha}\right)(R^\alpha_{\kappa,lm}) = 0$. In the evening, a Japanese arrives with a stack of paper and wants me to communicate to him my impressions at the peak of the hill!

27th. Excursion aboard Mitsui steamboat on China Sea. Walk through Shimonoseki, supper with Yamamoto & Watanabe. Afterward Nagai arrived. Poem with drawing for Mrs. Yamamoto and bedaubed much silk with inkbrush.[182] Most animated spirits. So much for peace and quiet!

28th. Rainy day. Evening invitation by the Commercial Club of Moji.[183] I play the fiddle, the Japanese

singen einzeln japanisch. Hört France. Schlau
und nicht so fein wie ~~Franz~~ Professoren, aber doch
mehr wie analoge Europäer. Auch hier Schlecht-
heit der Formen.

29. Rührender Abschied. Herr & Frau Yamamoto,
H. & F. Inagaki, Kuwaki (mit Söhnchen) Ishiwara,
Miyake sowie Herren der Mizzini- Gesellschaft. Alle
auf Schiff. Wunderbares Geschenk, Gedicht und Brief
von Zuckii (Sendai, Dichter) angekommen. Abfahrt
etwa 4 Uhr. Schiff gross und behaglich. Elektro-
dynamik Energietensor gefunden und Ishiwara
geschrieben.

30. Ruhige Fahrt. Gedanke für Ausgestaltung
von Weyl- 'Eddington' scher Theorie. Briefe
an Yamamoto und Zuckii. Lesen Frankfurter
Zeitungen, die Frau Prof. Berliner aus Tokyo
geschickt hat, trauriges Europa!

31. Ankunft in Shanghai bei herrlichem Wetter.
Mittags abgeholt von De Jong (Ingenieur) und
Herrn Gaston (Parvenü). ~~Abendessen~~ Wohnen bei letzterem.
Protz aber gutes Klavier. Sylvesterabend dort, ich sass
an seinem Klavier, sonst qualvoll und für mich traurig.
Shanghai unerfreulich. Europäer Musse chinesischer
Dienerschaft, sind faul, selbstbewusst und hohl.
Mittagessen bei de Jong. Freundlicher Engländer

sing one at a time in Japanese. High finance. Shrewd and not as refined as the ⟨Europ⟩ professors, but more like analogous Europeans, after all. Here, too, modesty of form.

29th. Touching farewell.[184] Mr. & Mrs. Yamamoto, Mr. & Mrs. Inagaki, Kuwaki (with little son), Ishiwara, Miyake, as well as gentlemen of the Mitsui company. All boarded ship.[185] Marvelous present, poem and letter from Zuckii (Sendai, poet) arrived.[186] Departure around 4 p.m.[187] Ship large and comfortable. Found electrodynamics energy tensor and wrote Ishiwara.[188]

30th. Peaceful trip. Idea for development of Weyl-Eddington theory.[189] Letters to Yamamoto and Zuchii. Read issues of *Frankfurter Zeitung* that Prof. Berliner's wife sent from Tokyo; sorrowful Europe![190]

31st. Arrival in Shanghai in glorious weather. At noon picked up by De Jong (engineer) and Mr. Gaton (parvenu). ⟨Supper⟩ Lodging with the latter vulgarian, but good piano. New Year's Eve there; I sat next to fine Viennese lady, otherwise noisy and, for me, doleful.[191]

⟨32nd⟩ 1st Jan. Shanghai unpleasant. Europeans (mass of Chinese servants) are lazy, self-assured, and hollow. Lunch at de Jong's. Friendly, internationally minded Englishman.

(in Gutons Haus)

von internationaler Gesinnung. Nachmittags, Reception,
jüdische und sonstige schmalzige Spiesser in
Schaaren, übliche Händedrücke und Reden –
abscheulich. Dann Diskussion im „Question-Klub"
(Komödie mit dummen Fragen). Abends noch
chinesisches Volks-Vergnügungs-Etablissement
besucht. Malerisches Leben. Chinese acceptiert wahllos
(Pełiminsfyery, Hochzeit, Beerdigung)
europäische Musik bei allen Gelegenheiten, gleich
ob Trauermarsch oder Walzer, wenn nur tüchtig
trompetet. Dort gab es auch mitten im weltlichen
Getriebe ein Tempelchen. Vormittags kleine Autofahrt
nach der Umgebung der Stadt; alles voll Grabhügel
und Särgen bezw. Sarghäuschen, die nicht entfernt
werden dürfen. Chinesen schmutzig, gequält,
stumpf, gutmütig, solid, sanft und – gesund.
Alle sind im Lobe des Chinesen einig, aber auch
über intellektuelle Minderwertigkeit bezüglich
Arbeit im Geschäft, bester Beweis: er erhält
10 mal weniger Lohn in der entsprechenden Stelle,
und der Europäer kann doch als Geschäfts-
angestellter erfolgreich mit ihm konkurrieren.
2. Januar. Mittags Abfahrt. Trübes, windiges Wetter.
Ich geniesse die Ruhe unbeschreiblich
3. Januar. kaltes, windiges Wetter. Beschauliches

In the afternoon, "reception" at Gaton's home, droves of Jewish and other schmaltzy, petty bourgeois, usual handshaking and speeches—abominable.[192] Then discussion at "Question Club" (comedy with stupid questions).[193] In the evening, also visited Chinese popular entertainment establishment. Picturesque life. Chinese indiscriminately accepts all European music for any occasion (merrymaking, wedding, funeral), never mind if funeral march or waltz, so long as there is plenty of trumpet blowing. There was also a small temple, in the midst of the worldly hustle and bustle. In the morning, short car ride to the city environs; everything full of burial mounds and coffins or small coffin houses, which cannot be removed. Chinese dirty, tormented, lethargic, good-natured, stable, gentle and—healthy. All are unanimous in praising the Chinese but also in regard to his intellectual inferiority in business skills; best evidence: he earns ten times lower wages in an equivalent position, and the European can still compete successfully with him as a business employee.

2nd January. Noon departure. Gloomy, windy weather. I enjoy the tranquility more than words can describe.

3rd January. Cold, windy weather. Contemplative,

beneidenswertes Dasein auf dem Schiff, wo
ich, um diesen Zustand zu erhalten, Bekanntschaften ängstlich vermeide. Nachdenken
und Rechnen über Eddingtons Theorie (Versuch
einer Vervollständigung derselben. Verbesserung der
Variationsmethode in der gewöhnlicher allgemeinen Relativität.
5. Januar. Ankunft in Hongkong 7 Uhr morgens.
Wir, hoffend einmal unsere Ruhe zu haben gehen
bummeln wir 9½ an Land, zunächst einiges zu
besorgen. Treffen bei Nippon Yusen Kaisha einen
Kerl, Namens Grün, der uns schon das letzte Mal Hongkong
gezeigt hatte. Er kündigte Empfang der Gemeinde
für Nachmittag an. Wir gingen noch zum französischen
Konsul und verabschiedeten uns schleunigst von
ihm. Wir fuhren wieder auf den Peak, ich stieg
noch hinauf auf den Gipfel. Prachtvolle Aussicht
auf Hafen, Meer und Insel. Ziemlich heiss
oben. Wir gingen zu Fuss in die Stadt herunter, etwa
eine Stunde abwärts, alles durch tropischen Hain.
Der ganze Weg war mit chinesischen Männern,
Frauen und Kindern voll, die störrend Ziegelsteine
nach oben schleppten. Ärmstes Volk der Erde,
grausam missbraucht und abgeschunden, schlimmer
als das Vieh behandelt. Lohn der Bescheidenheit
Sanftmut und Genügsamkeit. Dann aufs Schiff

enviable existence on board, where I, in order to maintain this state, fearfully avoid making acquaintances. Contemplation and calculations on Eddington's theory.[194] (Attempt at a completion of the latter.) Improvement of the variational method in regular general relativity.

⟨4⟩ 5th January. Arrival in Hong Kong. 7 a.m. In hope of being left in peace for once, we secretly go ashore at 9:30, initially to run a few errands. Met a fellow by the name of Gobin at Nippon Yusen Kaisha, who had already shown us Hong Kong the last time. He informed us of the [Jewish] community's reception that afternoon. We also went to see the French consul and quickly said our good-byes.[195] We drove up to the Peak again; I climbed up to the top. Glorious view of harbor, sea, and island. Quite hot up there. We walk on foot down to the town, downhill for about one hour, tropical groves all the way. The entire route was taken up by Chinese men, women, and children, groaning while hauling bricks uphill. Most pitiful of people on Earth, cruelly oppressed and abused, treated worse than cattle; their reward for modesty, gentleness, and frugality. Then drove to ship.

gefahren. Kaum abgekommen wieder von dem Kerl
abgeholt, ins jüdische Klubhaus neben Synagoge
gefahren. Trotz seiner emsigen Bemühung kann
sozusagen niemand zur "reception", was sehr komisch
war. Dann gingen wir noch notgedrungen zu
einer Familie zum Essen. Freitag-Abend-Gebet,
dann langes, furchtbar gewürztes Essen, ein
interessanter jüngerer russischer Jude dabei
Endlich heim - gottlob aufs Schiff.

6. Wundervoller wolkenloser frischer Morgen im
Hafen, 11 Uhr Abfahrt. Strahlende Sonne
liess Nachmittags lang in der Sonne, die gerade [mit Hut!]
noch ~~mit Hut~~ erträglich war. Fahrt durch Inselchen und zahlreiche chinesische Schiffe belebt, [Segel]
die auf den Wellen tanzen. Neue Idee für das elektro-
magnetische Problem der allg. Rel. Th.

7.+8. Trübe und feucht bei steigender Wärme
Nachdenken über allg. Relativität und Elektrizität

9. Niederschrift der Arbeit über Gravitation und
Elektrizität.

V. Briefe an Arrhenius, Planck, Bohr. Abends Ankunft vor Singapore
11. 6 Uhr Einlauf im Hafen. Drückende, trübe
Atmosphäre oft Regen. Monteur da mit Brief von
Java. Voote. Abtelegraphiert und geschrieben.

Barely arrived, again picked up by that fellow, driven to Jewish club-house beside synagogue. Despite his diligent efforts, practically nobody came to the "reception," which was very comical. Then we went, perforce, to his family['s] to eat. Friday evening prayers, then long, terribly spicy meal, an interesting young Russian Jew also there. Finally, thank God, back on board ship.

6th. Wonderfully cloudless crisp morning in the harbor. 11 a.m. departure. Brilliant sunshine. Sat for a long time in the sun in the afternoon, which (with hat) was still just barely tolerable. Voyage past islets and bustling with numerous Chinese sailboats dancing on the waves. New idea for the electromagnetic problem of the gen[eral] th[eory] of rel[ativity].

7th & 8th. Somber and humid with temperatures on the rise. Pondering gen[eral] relativity and electricity.

9th. Write down paper on gravitation and electricity.[196]

10th. Letters to Arrhenius, Planck, Bohr.[197] In the evening, mooring near Singapore.

11th. 6 o'clock, entry into harbor. Oppressive, gloomy atmosphere, frequent rain. Montor there with letter from Java, Voute. Declined by telegram and wrote [letter].[198]

Fahrt in konservierten Urwald bei leichtem
Regen. Unvergleichlicher Eindruck des wilden,
grossartigen Pflanzenwuchses, sumpfig, un-
durchdringlich. Mittagessen bei Montors.
Gegen Abend Fahrt zu Frankels Palmen-Kultur.
Bäume wundervoll, banale Menschen. Nachts
auf Schiff recht drückend.
12. Besuch bei Meyer-Löwens mit edler Tochter ("Perücke")
Dann bei dickem Weil mit fescher Gattin
richtige jüdische Zigeuner. Tropisches Haus mit
herrlicher Aussicht auf Stadt und Meer
Nachmittags bei strömendem Regen aufs
Schiff. Abfahrt 5 Uhr zwischen grellgrünen Sammet-Inselchen.
13. Ankunft in aller Frühe bei Malacca, wo das
Schiff auf offener See bis 3 Uhr nachmittags
hielt. Vormittags besuchten wir Malacca. Portu-
giesische Kirche u sonstige Bauwerke. Lebendiges
Gemisch von Indern, Malayen, Chinesen. Zweirädrige Wagen mit Strohdächern, gezogen von langgehörnten zigenden Rindern
Tropensonne, aber weniger feucht als in Singapore.
Dickes Haar in meiner Elektrizitätssuppe gefunden
am Nachmittag. Schade. Rechte Tropenhitze.
14. Mittags Ankunft in Penang. Bruthitze
auf dem Schiff, das ziemlich weit von der
Stadt in der Bucht liegen blieb. Menge Inder

Trip into jungle preserve in light rain. Unrivaled impression of wild, gorgeous plant growth, swampy, impenetrable. Midday meal at the Montors. Toward evening, trip to Fränkel's cultivated palms.[199] Trees wonderful, people banal. At night on board quite oppressive.

12th. Visit to Mayer-Croesus with noble daughter ("Portia").[200] Then to fat Weil with pretty wife, real Jewish gypsies. Palm house with splendid view of city and sea. In the afternoon, in pouring rain, back onto ship. Departure 5 p.m. between glaringly green velveteen islets.

13th. Arrival near Malacca in the early morning, where the ship held its position in the open sea until 3 p.m. We visited Malacca in the morning. Portuguese church & other buildings. Vibrant mix of Indians, Malays, Chinese. Two-wheeled carts with straw roofs, drawn by long-horned oxen. Blazing tropical sun but less humid than in Singapore. Discovered large fly in my electricity ointment in the afternoon. A pity. True tropical heat.

14th. Noon, arrival in Penang. Scorching heat on board the ship, which remained in the bay, quite far away from the city. A host of Indians

eingeladen, schöne hochgebaute Männer und Frauen. Wir fuhren um 3 Uhr an Land und liefen von Rickschas verfolgt in der Stadt herum, die sehr interessant ist. Schiffchen, Häuser, Menschen, hat alles Stil. Plätze in der Stadt recht ärgerlich. Wir sahen buddhistische Tempel mit geheimnisvoll-schreckhaften, farbigen Zierat, auch eine Moschee mit Bad, in der Männer lagen, eleganter arabischer Bau mit schlanken Formen von veredelter Farbe. Schönes, zudringliches Bettelweib. Mit Japanern zusammen in Nachen bei übel[...] Wetter aufs Schiff zurückgefahren. Else schwere Angst aber doch noch Kraft zum Schimpfen. Der kluge Inder mit flammenden schwarzen Augen [...] aber seelenruhig [...] und holte uns heil in der Harima ab. Bruthitze bis Mitternacht.

15. Fahrt bei angenehmer See. Neue Gedanken zum elektrischen Problem. Abends Gespräch über Sinn und Leben in Ceylon mit welschem Schullehrer. Herrlicher Sternenhimmel. Beneidenswerte Existenz.

16.–18. Strenge Arbeit am Problem trotz Hitze. Geht vorwärts mit vielen Rückschlägen.

19. Colombo. Morgens vergebliche Bemühung, Wagenpartie

were invited on board, handsome, tall-statured men and women. We went ashore around 3 p.m. and walked, pursued by rickshaws, around town, which is very interesting. Small boats, houses, people, they all have style. Heat downtown quite tolerable. We saw Buddhist temples with mysterious, terrifying, colorful decor, also a mosque with baths inside where the men were lounging about, elegant Arab structure with slender towers, whitish in color. Beautiful, intrusive beggar woman. Traveled back to ship in dinghy in rolling waves, together with Japanese. Else very frightened but had sufficient energy left to grumble. But the gaunt Indian with flaming black eyes calmly rowed his long oar steadily on and delivered us safe and sound to the Haruna. Sweltering heat until midnight.

15th. Cruising with pleasant breeze. New ideas about the electrical problem. In the evening, conversation with Indian school teacher about land and life in Ceylon. Splendid star-studded night sky. Enviable existence.

16th–18th. Rigorous work on the problem despite the heat. Moving forward with many setbacks.

19th. Colombo. In the morning, futile attempt to assemble a car party.

zusammen zu bringen. Interessante Feststellung über japanische Geschäftspraxis von Kapitän, der draht-
lose Telegramme von billiger offerierenden Ver-
kehrsformen aus Colombo aus bucht zu
erratenden Gründen dem Passagieren verschweigt.
Vormittags Spazierfahrt im Gang Bahnhof tort
in Begleitung zudringlicher Eingeborener.
1ʰ25 Fahrt nach Negombo, Etwas nördlich
an Küste gelegenes Städtchen ohne Europäer.
Wir nahmen 2 Rickscha-Männer, deren einer
ein absolut nackter Naturmensch, der andere
ein früherer Hagenbeck-Elefanten-Pfleger
war, der von Hamburg schwärmte. Sie fuhren
uns durch das Städtchen, dessen Haupt-
Häuserreihe wie einzelne Häuser in
Palmenhaine geduckt sind. Wir werden
überall begafft wie bei uns Singhalesen.
Dann Besuch von Fischerdorf. Kinder ganz
nackt, Männer Lendentuch. Schöne Leute
Fischerboote aus zwei starr verbundenen
schmalen parallelen Teilen. Grosse
Schnelligkeit aber unbequemer Sitz
Boot kam heim mit viel Fischen
und einem Schwarm neidischer Raben.

Interesting conclusion about ⟨Japanese⟩ captain's business practice, who, for easily surmised reasons, keeps passengers in the dark
about cheaper wireless telegrams offered by communication firms
from Colombo. In the morning, excursion by tram & on foot, train
station, fort, in the company of intrusive natives. 1:25–3:15 p.m., trip
to Negombo, a small town with no Europeans somewhat northward
along the coast.[201] We hired two rickshaw men, one of whom was a
completely naked primitive man, the other a former Hagenbeck Zoo
elephant-keeper, who enthused about Hamburg.[202] They drove us
through the little town; its main terraces are like detached houses
tucked under palm groves. We were ogled at everywhere, just as the
Sinhalese are back home. Then, visit to a fishing village. Children
stark naked, men loincloth. Handsome people. Fishing boats made
of two stiffly attached, narrow parallel pieces. Great speed but uncomfortable sitting.

Boat returned home with many fish and a swarm of envious crows.

Dann an Bucht und Flüsschen vorbei
vor ihm in der Wiese sahen wir auf 12 m
Distanz grosses Krokodil, dass ich
nach Steinwürfen und Schreien vieler
Eingeborener ganz langsam ins Wasser
trollte. Dann europäisches Gasthaus und
Bahn. Engländer führen tadellose Verwaltung ohne unnötige Chikanen. Ich hörte
von keinem Worte der Unzufriedenheit
gegen sie, auch nicht von einem singhalesischen Lehrer,
der als Zwischendeck-Passagier von
Penang nach Colombo mitfuhr. Der Rick-
scha-Mann war von uns bezw. unseren
5 Rupien so begeistert, dass er uns zum
Abschied Bananen an die Bahn brachten.
Am Bahnhof lernten wir noch eine
bildschöne junge Singhalesin nebst
Schwester und Mutter kennen, Dorfaristokraten.
Sie hatte aber einen holländischen
Urgrossvater. Selten hab ich so etwas
Feines gesehen. Besonders prächtiger
Sternhimmel. Heimfahrt noch immer
heiss und Gewimmel von Mücken im Wagen,
was in dieser Reis-Gegend recht

Then, past bay and rivulet, in front of which we saw a large crocodile at a distance of 12 m, lying in the grass; after being pelted with stones and screamed at by many natives, it slid into the water, very slowly. Then, European restaurant and train. The English govern impeccably without needless chicanery. I did not hear words of discontent against them from anyone, not even from a Sinhalese teacher who traveled along as a steerage passenger from Penang to Colombo. The rickshaw man was so delighted by us, or rather by our five rupees, that he brought bananas to us at the train station in parting. At the station we also made the acquaintance of a ravishingly beautiful, fine young Sinhalese woman with sister and mother; village aristocrats. But she had a Dutch great-grandfather. Rarely have I ever seen anything so fine. Particularly magnificent starry sky. Homeward trip still hot and teeming with mosquitoes in the carriage, which in this ⟨swampy⟩ rice region was

unheimlich war. In Colombo angekommen
fielen die Rickschw-kulis über uns
her. Nach längeren vergeblichem Wider-
stand ergaben wir uns ihnen. Sie betrachten
es als eine Unverschämtheit von einem
Europäer, wenn er zu Fuss geht. Dann
Fahrt aufs Schiff bei hohem Seegang
im Kinderschiff unter Todesangst und
schweren Vorwürfen Elses. Nachher Löschen
des furchtbaren Durstes auf dem Schiff,
Nachts fast keine Abkühlung. Temp. tags etwa
29 Grad im Dampfer im gut ventilierten Teil,
ausnahmsweise kühl für diese Gegend.

22. Arbeit über Gravitation und Elektrizität endgültig
aufgeschrieben. Seereise überaus angenehm ohne
bemerkenswerte Erlebnisse. Typensches Essen
mit Kapitän. Wundervolle sternklare Abende Temperatur
sinkt langsam mit Entfernung vom Aequator.
Telegramme über Einwirken der Franzosen ins Ruhr-
gebiet, sind in 100 Jahren nicht gescheiter geworden.

31. Im roten Meer zuerst Konstant 28-29 Grad
bei fast klarem Himmel. Wunderbare
Sonnenuntergänge mit gelbroten bis purpur-
roten Himmel und grell beleuchteten

quite eerie. Upon arrival back in Colombo the rickshaw coolies pounced on us. We surrendered to them after lengthy, futile resistance. They regard it as the height of effrontery if a European travels on foot. Then, trip by rowboat to ship in high swell with Else's deadly terror and severe reproaches. Afterward on board, quenching of dreadful thirst. At night almost no cooling down. Daytime temp[erature] about 29 degrees on the steamship in the well-ventilated section; unusually cool for this region.

22nd. Final draft of paper on gravitation and electricity.[203] Sea voyage exceedingly pleasant without noteworthy experiences. Japanese meal with captain. Wonderful starlit evenings. Temperature drops slowly with distance from the equator. Telegrams about the French marching into the Ruhr region, have not grown any wiser in 100 years.[204]

31st. On the Red Sea, first a constant 28–29 degrees with almost clear skies. Fabulous sunsets with yellowish red to purplish red sky and sharply jagged islets, glaringly lit up bezw. silhouettenhaft dunklen scharfgezackten Inselchen. Heute Ankunft in Suez. Tiefblaues auffallend durchsichtiges Meer Himmel gleichmässig schwach bedeckt. Silbrige matte Farben, malerische Segelschiffe, gelbe Ufer — Fahrt durch Bitterseen mit öden verlassenen Ufern wundervoll. Matte silbrige Töne. Himmel trübe, Luft ziemlich kühl.

Am letzten heissen Tag war Maskenfest der Passagiere, tags vorher der Stewards. Japaner und Portugiesen in diesem Kunst. Nette Bekanntschaften in der letzten Teil gemacht. Griechischer Gesandter, der aus Japan heimkehrt, sympathische englische Natur, der ein Pfund für die Jer. Universität spendet trotz meiner Proteste, nicht zu vergessen das Ehepaar Ogata, ferner liebenswürdige japanische Kaufleute, mit denen wir viel plauderten, ~~auf dem Schiff.~~

1. Februar Ankunft in Port Said in aller Früh. Griechischer Gesandter erleichterte uns die Landung u. Vorstellung. Junger Jude Goldstein erscheint beim Zollhaus mit Telegramm von Jerusalem, um uns zu helfen. Stadt

or in dark silhouette. Today, arrival in Suez. Deeply blue, remarkably transparent sea. Sky mildly veiled in mist all over. Silvery dull colors, picturesque sailboats, yellow shoreline. Passage through saline lakes with barren, abandoned shores. Dull silvery hues, sky gloomy. Air quite cool.

On the last hot day there was a masquerade party for the passengers, the day before, for the stewards. Japanese are virtuosos of this art. Recently made some nice acquaintances. Greek envoy, who is returning home from Japan, likeable English widow, who contributes one pound for the Jer[usalem] university despite my protests, not to forget the Og[a]tas—fine, pleasant Japanese merchants, with whom we chatted a lot on board the ship.

1st February. Arrival in Port Said in the early hours. Greek envoy facilitated our landing and customs. Young Jew ⟨Goldstein⟩ Cantor appears at customs house with telegram from Jerusalem to assist us. City a

wichtiger Fremdenplatz mit entsprechendem
Gesindel. Besuch beim Gemeindevorstand
(Palästinenser). 6 Uhr abends Zug nach Landara
am Suezkanal. Canton und sein Compagnon
Goldstein begleiten uns dorthin und auf Fähre
über den Kanal von 8 – 1 Uhr nachts Aufenthalt
Dann Abfahrt nach Palästina, erleichtert
durch jungen jüdischen Konducteur, der
mich in Berlin in Versammlung gesehen
hatte – von Mitjuden nicht restlos begeistert,
aber braver, guter Mensch. Fahrt zuerst durch
Wüste, dann von etwa 7½ Uhr ab
durch Palästina bei ziemlich trübem Wetter
und häufigem Regen
Fahrt zuerst durch sehr spärlich bewachsene
Ebene, durch Araberdörfer und jüdische
Kolonien unterbrochen, Ölbäume, Kakteen
Orangenbäume. An Zweigstation unweit Jerusalem
von Assuschkin, Mossensohn und einigen andern
der unsrigen empfangen. Fahrt an Kolonien vorbei
durch wunderbares Felsthal hinauf nach Jerusalem.
Dort Ginzberg, hohes Klaraschen. Im Auto mit Offizier
zum Schloss des Gouverneurs, ehemals im Besitz
Kaiser Wilhelm's, ganz wilhelminisch. Herbert

real gathering place of foreigners with corresponding riff-raff. Visit to the mayor (Palestinian). 6 p.m., train to Qantarah at the Suez Canal.[205] Cantor and his companion Goldstein accompany us there and by ferry across the canal, stopover from 8–1 a.m. at night. [2nd.] Then departure to Palestine facilitated[206] by young Jewish conductor, who had seen me in Berlin at a public rally—not completely thrilled with his fellow Jews, but a reliable, good person. Trip first through desert, then from about 7:30 through Palestine in quite dreary weather and frequent showers.

Trip first through flat terrain with very sparse vegetation, interspersed with Arab villages and Jewish colonies, olive trees, cactuses, orange trees. At a junction not far from Jerusalem, welcomed by Ussishkin, Mossinson, and a few others of our [fellow Jews].[207] Travel past colonies through wonderful rock valley up to Jerusalem.[208] There Ginzberg, happy reunion.[209] In automobile with officer to governor's palace, formerly a possession of Kaiser Wilhelm, very Wilhelminian. Make acquaintance with Herbert

Samuel kennen gelernt. Englische Form. Hohe viel-
seitige Bildung. Hohe Lebensauffassung, mit Humor
gemildert. Schlichter feiner Sohn, frohmütige,
derbe Schwiegertochter mit nettem Söhnchen.
Tag regnerisch, aber doch Ahnung von der
herrlichen Aussicht auf Stadt, Hügel,
Totes Meer und transjordanische Berge.
3. Mit S. Samuel zu Fuss in die Stadt auf (Subbat!)
Fussweg an Stadtmauer vorbei zu
malerischem altem Thor. Weg in die Stadt
bei Sonnenschein. Strenge kahle Hügelland-
schaft mit weissen, oft gekuppelten weissen Stein-Häusern
und blauem Himmel herreissend schön,
ebenso die in quadratische Mauer gedrängte
Stadt. Wester und Ginzberg in die Stadt.
Durch Bazar - Gassen und sonstige
enge Gassen zur grossen Moschee auf
herrlichem weitem erhöhtem Platze, wo Salomos
Tempel stand. Aehnlich Byzantinischer Kirche
polygonal mit mittlerer, von Säulen getragener
Kuppel. Auf anderer Seite des Platzes Basilika-
ähnliche Moschee von mittelmässigem Geschmack.
Dann hinunter zu Tempelmauer (Klagemauer), wo
strumpfbeinige Stammesbrüder laut beteten,

Samuel. English formality. Superior, multifaceted education. Lofty view of life tempered by humor. Unassuming, fine son, cheerful, earthy daughter-in-law with nice little son.[210] Day rainy, yet hint of the magnificent view of city, hills, Dead Sea, and Transjordanian mountains.

3rd. Walked with S[ir] Samuel into the city (Sabbath!) on footpath past the city walls to picturesque old gate, walk into the city in sunshine. Stern bald hilly landscape with white, often domed white stone houses and blue sky, stunningly beautiful, likewise the city pressed inside the square walls.[211] Continue on into the city with Ginzberg. Through bazaar alleyways and other narrow streets to the large mosque on a splendid wide raised square, where Solomon's temple stood. Similar to Byzantine church, polygonal with central dome supported by pillars.[212] On the other side of the square, a basilica-like mosque of mediocre taste.[213] Then downward to the temple wall (Wailing Wall), where obtuse ethnic brethren pray loudly,

22. The Einsteins at Government House, Jerusalem, 6 February 1923. Front row: Edwin Samuel, Elsa Einstein, Sir Herbert Samuel, Lady Beatrice Samuel, Einstein, and Édouard-Paul Dhorme. Middle row: Antonin-Gilbert Sertillanges and Margaret Richmond-Lubbock. Back row: Norman Bentwich, Ernest T. Richmond, and Gaudenzio Orfali (courtesy of the École Biblique, Jerusalem).

mit dem Gesicht der Mauer zugewandt,
den Körper in wiegender Bewegung vor und zurück
bewegend. Klagevoller Anblick von Menschen mit
Vergangenheit ohne Gegenwart. Dann Diagonal
durch die (sehr dreckige) Stadt, die von verschie-
densten Heiligen und Rassen wimmelt,
geräuschvoll und orientalisch-fremdartig. Prächt-
voller Spaziergang über den zugänglichen Teil
der Stadtmauer. Tango zu Ginzberg – Ruppens
zum Mittagessen mit fröhlichen und ernsten
Gesprächen. Aufenthalt wegen starken Regens.
Besuch im Bucharischen Judenviertel und
in düsterer Synagoge, wo gläubige, schmutzige
Juden betend Sabbatende erwarten. Besuch
bei Bergmann, dem ersten Prager Heiligen,
der die Bibliothek einrichtet mit zu wenig
Raum und Geld. Fürchterlicher Regen mit
vermehrtem Strassendreck. Heimfahrt mit Ginzberg
und Bergmann im Auto.
4. Mit Ginzberg und J. Samuels tüchtiger
und derbnatürlicher, froher Schwiegertochter über
wunderwoll kahle weiche Hügel und eingeschnittenen
Thälern im Auto bei strahlender Sonne und frischem
Wind im Auto hinunter nach Jericho und zu

with their faces turned to the wall, bend their bodies to and fro in a swaying motion.[214] Pitiful sight of people with a past but without a present. Then diagonally through the (very dirty) city swarming with the most disparate assortment of holy men and races, noisy, and Oriental-exotic. Gorgeous walk over the accessible part of the city ramparts.[215] Then to Ginzberg-Ruppin for midday meal with cheerful and serious conversations.[216] Detained due to heavy rain. Visit to the Bukharian Jewish quarter and to gloomy synagogue where pious, grimy Jews await the end of the Sabbath in prayer.[217] Visit Bergmann, the serious holy man from Prague who is setting up the library with too little space and money.[218] Atrocious rain with more and more street filth. Homeward drive with Ginzberg and Bergmann by car.

4th. Drive with Ginzberg and S[ir] Samuel's capable and earthy cheerful daughter-in-law[219] over wonderful bleak rolling hills and carved valleys in beaming sunshine and a fresh wind, downhill to Jericho and

dessen alten Ruinen. Herrliche tropische
Oase in Wüstengegend. Essen in Jericho-Hotel.
Dann Fahrt durch breites Jordanthal bis
Jordanbrücke auf mygchener morastiger
Strasse, wo wir prachtvolle Beduinen
sahen. Dann in glänzender Sonne wieder
heim. Inhause *bei scheidender Sonne* prächtiger Blick auf
totes Meer und transjordanische Hügel
von Amtswohnung von S. Diez aus, wo
wir The nahmen. Dann im dunkel
werdenden Zimmer interessantes Gespräch
über Religion, Nationalität mit S. Diez, Abends
*gemütliches* Gespräch mit S. Samuel u. Schweizertochter.
Tag von unvergesslicher Pracht, Einzigartiger
Zauber dieser strengen monumentalen Natur
mit ihren dunklen *eleganten* arabischen Söhnen *in ihren Lumpen*. Hierbei—
nige Kamele und Esel gabs viele zu sehen.

5. Besuch zweier *jüdischen* Baukolonien westlich Jerusalem,
zur Stadt gehörig. Der Bau wird ausgeführt
von jüdischer Arbeitergenossenschaft, in
welcher die Leitenden Personen gewählt
werden. Die Arbeiter kommen ohne Fach-
kenntnisse und Übung an und leisten
nach kurzer Zeit vortreffliches. Die Leitenden

its ancient ruins. Splendid tropical oasis in desert region. Meal at Jericho Hotel. Then drive through wide Jordan Valley up to the Jordan bridge on horrendously marshy road, where we saw magnificent Bedouins.[220] Then in dazzling sunshine homeward again. At home, in setting sun, resplendent view of Dead Sea and Transjordanian hills from S[ir] Diez's official residence, where we had tea.[221] Then in the darkening room, interesting conversation about religion and nationality with S[ir] Diez. In the evening, pleasant chat with S[ir] Samuel & daughter-in-law. A day of unforgettable magnificence. Extraordinary enchantment of this severe, monumental landscape with its dark, elegant Arabian sons in their rags. There were many four-legged camels and donkeys to be seen.

5th. Tour of two Jewish colonial settlements west of Jerusalem, belonging to the city. The construction is carried out by Jewish workers' cooperative, in which the leaders are elected. The workers arrive without expertise or training and achieve excellence after a short time. The leaders

erhalten nicht mehr Gehalt als die
Arbeiter. Besuch der jüdischen Bibliothek
Bergmann aus Prag schaltet dort tüchtig
aber humorlos. Ein hiesiger Mathematiker
(Gymnasiallehrer) zeigt mir manches über
seine wirklich interessanten Untersuchungen
betreffend stetige Matrices und ihre
Operationen. Abends musiziert mit Offizier
in I. Samuels Wohnung, viel zu lang, weil
ausgehungert nach Musik.

5. Besuch in jüdischer Kunstschule. Tüchtige
Arbeit unter schweren Verhältnissen. Wieder-
belebung antiker jüdischer Ornamente. Nachmittags Begrüssungsempfang durch jüdische
Schüler, die Spalier bildeten Feierliche und
darauf in Schulsaal von jüdischen Bürgern
überhaupt. Ansprachen von Ussischkin und
Yellin und Überreichung von hebräischer
Adresse. Abends musikalische Einladung
bei Bentwitsch. Höchst musikalische Familie.
Wir spielten Mozartquintett.

6. Grabkirche. Via Dolorosa. Nachmittags Vortrag
(französisch) im Universitätsgebäude an
spre. Ich muss mit hebräischer Begrüssung

23. Einstein and Elsa inscribed in the "Golden Book" of the Jewish National Fund, 6 February 1923 (with permission of the Albert Einstein Archives, the Hebrew University of Jerusalem).

are paid no more than the workers.[222] Visit to the Jewish library. Bergmann from Prague operates there ably but with no sense of humor.[223] A local mathematician (high-school teacher) shows me some of his really interesting investigations regarding continuous matrices and their operations.[224] In the evening, music making with officer in S[ir] Samuel's apartment, for far too long, because starved of music.[225]

5th. [6th.] Visit of Jewish school of art. Splendid work under difficult conditions. Revival of antique Jewish ornaments.[226] In the afternoon, welcoming reception by Jewish schoolchildren forming a guard of honor and then by Jewish citizens in general in school hall. Speeches by Ussishkin and Yellin and presentation of Hebrew address.[227] In the evening, invited for music at Bentwitch's. Highly musical family. We played Mozart quintet.[228]

6th. [7th.] Church of the Holy Sepulcher. Via Dolorosa.[229] In the afternoon, lecture (French) in university building *in spe*. I have to commence with greeting in Hebrew,

beginnen, die sich mühsam ablesen. Nachher
Dankworte (recht witzig) von Hrn. Samuel
& Spaziergang über die Bergstrasse hin und
her. Philosophische Gespräche. Abends grosse
Honoratioren - Einladung mit gelehrten und
anderen Gesprächen. Abends von all diesen
Komödien restlos befriedigt!
8. Fahrt nach Telaviw 9½ –12 im Auto.
Empfang im Gymnasium. Besuch mehrerer
Stunden. Freiübungen der Schüler. Kurzer Dank
an diese. Empfang im Rathaus. Zum Ehrenbürger
ernannt. Rührende Ansprache. Nach dem Mittagessen
Besuch der im Bau begriffenen elektrischen
Rutenberg – Centrale, der städtischen Kraftzentrale,
des Quarantäne – Lagers, der Sand – Bausteine –
Fabrik. Dann grosse Volks – Begrüssung vor
dem Gymnasium mit Ansprache von Mossinson
und mir. Besuch der landwirtschaftlichen
Versuchsanstalt, der wissenschaftlichen Abend-
kurse (Czerniawski) und des Ingenieurvereins,
wo ich ein Diplom und eine prächtige silberne
Dose erhielt. Abendessen bei Tolkowski. Abends
Versammlung der Gebildeten mit Rede von
mir. Die Tätigkeit der Juden in manigem

24. Reception for Einstein at Tel Aviv city hall, 8 February 1923. Front row includes Elsa and Albert
Einstein and Mayor Meir Dizengoff (to Einstein's left). Back row includes Ahad Ha'am and Benzion
Mossinson and other members of the Tel Aviv city council (courtesy of the Central Zionist Archives,
Jerusalem).

which I read out cumbersomely. Afterwards, words of thanks (quite witty) by Herb[ert] Samuel & stroll back and forth on the hill road.[230] Philosophical conversations. In the evening, large reception attended by dignitaries with scholarly and other conversations.[231] That evening, well and truly satisfied with all these comedies!

⟨7⟩ 8th. Drive to Tel Aviv 9:30–12 in automobile. Reception at high school. Visit lasted several hours. Calisthenics by pupils. Short thanks to them.[232] Reception at city hall; named honorary citizen. Moving speech.[233] After lunch, visit to the Ruthenberg electrical center under construction, the municipal central power station, the quarantine camp, the sand-brick factory.[234] Then large public welcome in front of the high school with speeches by Mossinson and me,[235] visit of the agricultural experiment station, the scientific night courses (Czerniawski) and the engineers' association, where I received a diploma and a splendid silver container.[236] Supper at Talkowski's. In the evening, gathering of scholars with speech by me.[237] The accomplishments by the Jews in but a few

Juden in dieser Stadt erregt die höchste
Bewunderung. Moderne hebräische Stadt
aus dem Boden gestampft mit regem wirt-
schaftlichem und geistigen Leben. Ein
unglaublich reges Volk, unsere Juden!
Besuch landwirtschaftlichen Schule. Mikwe
und der jüdischen Rotschildkolonie Rischon
le Zion. Beide ~Anstalt~ Werke schon 50
Jahre alt. Alter Mann hielt auf dem Dorf
Begrüssungsrede. Schulstunde, Kinder
im Garten. Froher Eindruck von gesundem
Leben, aber wirtschaftlich doch nicht recht
selbständig. Eisenbahnfahrt nach Jaffa
durch Ebene mit allmählich näher rückenden
Gebirge. Arabische und jüdische Siedelungen
Jüdische Salzanlage Station vor Jaffa. Arbeiter
kamen zur Bahnhof und grüssten uns
Ankunft nach Sabbatbeginn in Haifa trotz
Herrn Strucks vorheriger Warnung. Gang zu
Fuss durch wenigen Dreck mit Ginzberg zu
dessen Schwager Pfizner. Frau zart und von
scharfer Intelligenz. Gemütliches Zimmer
oben, Deutsche Dienstmädchen. Abends Menge
Leute aus Neugrade, aber auch Struck und

25. Reception for the Einsteins at the colony of Rishon LeZion, 9 February 1923. Left to right: Menashe Meirowitz, Avraham Brill, Aharon Eisenberg, Einstein, Avraham Dov Lubman-Haviv, Elsa Einstein, Rivka Abulafia, and Miriam Meirowitz (courtesy of Kedem Auction House Ltd.).

years in this city excite the highest admiration. Modern Hebrew city stamped out of the ground with lively economic and intellectual life. What an incredibly lively people our Jews are!

⟨8⟩ 9th. In the morning, labor assembly. Very impressive.[238] Visit to the agricultural school. Mikve Large wine cellars. Eggs being artificially incubated must be cooled once a day.[239] And the Jewish Rothschild colony, Rishon le Zion.[240] Both enterprises already 50 years old. Old man delivered a welcome speech in the village. School lesson, children in the garden. Cheerful impression of healthy life, but economically still not yet quite independent. Train ride to Jaffa[241] together with Joffe (physician and cousin of the Russian) through plains with gradually approaching mountains. Arab and Jewish settlements. Jewish salt works station before Jaffa. Workers came to train station to greet me.[242] Arrival in Haifa after the beginning of the Sabbath, despite Mr. Struck's prior warning. Walk through enormous filth with Ginzberg and the physicist Czerniawski to his brother-in-law's, Pefzner. Wife delicate and of acute intelligence.[243] Cozy room upstairs. German maids. In the evening, a lot of indifferent people [present] out of curiosity, but also Struck and

Frau

10. Realschule besichtigt (Sabbath). Verpreusster
aber tüchtiger Direktor Biram. Strucks Wohnung.
Mittagessen bei ihm mit gemütlichen Gesprächen
Besuch bei Neumanns Mutter, umgeben
von X Söhnen, Töchtern etc. Spaziergang
auf den Karmel mit Struck. Jüdische —
Arbeiter angetroffen. Auf Pastors Dach gestiegen,
mit prachtvollem Rundblick auf Haifa
und Meer. Jüdischer Chaluz begleitet uns schräge
Strasse hinab bis zu Wohnung eines arabischen
Freundes. Das kleine Volk kennt keinen Nationalismus. *Besuch eines arab. Schriftstellers mit der Tochter Frau*
Abends Feier im Technikum. Wieder Reden,
bemerkenswerte von Tschernichowski und Auerbach.
Psalmen und ostjüdische Lieder des Kozen-
schen.
11. Besuch der Werkstätten des Technikums.
Dann Rothschild-Mühle und Ölfabrik
Erstere fast fertig. Ungeheuer raffinierte —
fast ganz automatische Einrichtung.
Nachmittags Fahrt über Ebene Israel, Bethlehem *Nazareth*
nach Tiberiassee. Unterwegs Besuch der im Bau
begriffenen Kolonie Nahallal, die nach Kauffmanns
Plänen gebaut ward. Fast alles Russen.

wife.[244]

10th. Viewed the Reali high school (Sabbath). Prussified but capable director Biram.[245] Struck's apartment. Lunch with him there with pleasant conversations. Visit to Weizmann's mother, surrounded by umpteen sons, daughters, etc.[246] Walk up [Mt.] Carmel with Struck. Encountered Jewish woman laborer. Climbed onto pastor's roof with magnificent panorama of Haifa and sea.[247] A Jewish *Chaluz* [pioneer] accompanied us down the sloping street to the apartment of an Arab friend. The common folk know no nationalism. Visit to an Arab writer with German wife.[248] In the evening, reception at the technical college. Again speeches, remarkable ones by Czerniawski and Auerbach. Psalms and Eastern European Jewish songs by candlelight.[249]

11th. Visit to the technical college workshops. Then Rothschild mill and oil factory. The former almost finished. Incredibly sophisticated, almost entirely automated installation.[250] Afternoon trip across the Jezreel Valley, ⟨Bethlehem⟩ Nazareth, to Lake Tiberias.[251] On the way, visit to the Nahalal colony, which is being constructed according to Kauffmann's design. Virtually all Russians.

Dorf mit privaten Parzellen. Bauerbest genossenschaftlich. Nach Nazareth, zu dem man nach malerischer Bergfahrt gelangt, strömender Regen. Fahrt in die Nacht hinein bis zu dem Gut Migdal. Das letzte Stück muss Auto von Maul- Tieren durch grossen Dreck bis zum Guts- haus gezogen werden. Gemütlicher Abendhock mit opulentem Mahl. Gut soll in Gartenparzellen aufgeteilt werden. Unser Wirt ist ein strammer, sesshaft gewordener Zigeuner. Seine Familie hat's dort nicht ausgehalten und lebt gegenwärtig in Deutschland. Drollige Pilgerfahrten mit grosser Laterne nach dem Abtritt. Strömender Regen in der Nacht.
12. Spazierfahrt hinunter zum Tiberiassee. Palmen- und Pinien-Allee. Landschaft ähnelt Genfersee. Sonne kommt hervor. Üppige aber Malaria-verseuchte Gegend. Bildhübsche junge Mädchen und interessanter studierter Arbeiter im Gut/Hause. Nach Mittagessen über das malerische Tiberias nach kommunistischer Siedlung Degania beim Abfluss des Jordans

Village with private allotments. Construction work cooperative.[252] To Nazareth, reachable after a scenic mountain drive, in pouring rain. Drive into the night up to the Migdal farm. For the last leg, the car had to be dragged through deep mud by mules up to the Migdal farmhouse. Cozy evening get-together with opulent meal. Farm is supposed to be divided up into garden plots. Our host is a strapping former gypsy who has put down his roots. His family was unable to stand it there and is presently living in Germany.[253] Hilarious pilgrimages with a big lantern to the outhouse. Torrential rain during the night.

12th. Car ride down to Lake Tiberias. Palm and pine avenue. Landscape resembles Lake Geneva. Sun emerges. Lush region but malaria infested. Ravishingly beautiful young Jewess and interesting educated worker in the farmhouse. After lunch via picturesque Tiberias to the Communist settlement Degania at the effluent of the Jordan

aus Tiberias-See, vorher Megdala passiert,
Marias Heimatsort, wo Araber Archäologen
Boden zu horrenden Preisen verkauften.
Kolonisten höchst sympathisch, meist
Russen. Schmutzig aber von erster Wollen
und mit Tüchtigkeit und Liebe ihr Ideal
fortsetzend den Kampf gegen Malaria,
Hunger und Schulden. Dieser Kommunismus
wird nicht ewig dauern aber ganze Menschen
erziehen. Nach eingehender Unterhaltung
mit Besichtigung Fahrt bei gutem Wetter
nach Nazareth hinauf. Unterwegs
prachtvolle Aussicht auf See, felsige
Hügel und zuletzt das malerische
Städtchen Nazareth. Abends in deutschem
Gasthaus, heimelig. Strömender Regen
aufs Neue.
13. Autofahrt von dem terrassenartig gebauten, sehr
malerischen Nazareth über Jesreel, Nablus
nach Jerusalem. Abfahrt bei ziemlicher Hitze,
dann empfindlich kalt mit strömendem
Regen. Unterwegs Strasse durch eingesunkenes
Lastauto versperrt. Menschen und Auto separat
Umweg über Graben und Feld. Die Autos werden

from Lake Tiberias, first passing by Magdala, Mary's home town, where Arabs sold land to archaeologists at horrendous prices.[254] Colonists extremely likable, mostly Russians. Grubby but of earnest will and with tenacity and love in pursuit of their ideal, carrying on the fight against malaria, hunger, and debt. This communism will not last forever but will raise people of integrity. After detailed conversation and tour, drive up to Nazareth in good weather. Along the way, magnificent view of the lake, rocky hills, and finally the picturesque little town of Nazareth. In the evening at a German inn, homelike.[255] Pouring rain anew.

13th. Drive from terraced, very scenic Nazareth across the Jezreel Valley, Nablus, to Jerusalem. Quite hot at departure, then severe cold with pelting rain. En route, road blocked by a truck sunk in the mud. People and car take separate detours over ditch and field. Cars get

schwer geschunden in diesem Lande. Abends
deutscher Vortrag in Jerusalem in vollgestopf-
tem Saale mit unmenschlichen Ansprachen
und Diplom der Palä jüdischen Ärzte, bei
dessen Überreichung der Redner Angst hatte
und stecken blieb. Gottlob, dass es auch
unter uns Juden weniger selbstbewusste
gibt. Man will mich unbedingt in Jerusalem
haben und attackiert mich in diesem Sinne
in geschlossener Reihe. Das Herz sagt ja, aber
der Verstand nein. Meine ~~Fräulein~~ Else Vorschied
der Reise in schwerem Fieber.

14. 6¾ Abfahrt mit Hadassa nach Bahnhof.
7½ Abfahrt und Abschied an der Bahn
Hadassa fährt mit bis Lud. Umsteigen.
Frau immer schlechter bis Candara, wo
sie ganz zusammenbricht. Freundlicher
arabischer Kondukteur. In Candara Bekanntschaft einiger Beamten, die meiner Frau
Esser und Lager gewähren 5½ – 10 Uhr Aufenthalt. Weiterreise sehr schwer. Ankunft
in Port Said. Bei Herrn Muschli in schönem
Haus geborgen. Alles wird gut werden.

15. Spaziergang zum Lesseps-Denkmal. Kubistisches

heavily battered about in this country. In the evening, lecture in German in Jerusalem in a packed hall with inevitable speeches and presentation of diploma by Jewish medical doctors, the speaker scared stiff and froze.[256] Thank heavens that there are also some with less self-assurance among us Jews. I am wanted in Jerusalem at all costs and am being assailed on all fronts in this regard. My heart says yes but my mind says no. ⟨My wife⟩ Else seriously feverish on the eve of our voyage.[257]

14th. 6:45 departure with Hadassa to railway station. 7:30 departure after farewells by the tracks. Hadassa travels with us until Lod.[258] Change of trains. Wife feeling increasingly worse up to Qantarah, where she completely collapses. Kind Arab conductor. In Qantarah, acquaintance with a few officials who give my wife eggs and a place to lie down. Layover 5:30–10 p.m. Onward travel very arduous. Arrival in Port Said. Safely housed in Mr. Muschli's beautiful house.[259] All will be well.

15th. Walk to Lesseps monument.[260] Cubist

26. The Einsteins at Port Said with Max Mouschly and Celia Mouschly-Turkel, 16 February 1923 (with permission of the Albert Einstein Archives, the Hebrew University of Jerusalem).

## 39

(Bild von viereckigen Badhäuschen, und

grössern Häusern längs Strand. Strahlende Sonne. Gefühl von

Befreiung. Frau besser, wird aufopfernd von Frau Muselli

gepflegt. Besuch bei Governo (Orientale mit blauem Gesicht) und einigen Konsuln.

16. Vormittag Abreise auf „Oronz" (Oriental line)

Schlechtes Futter. Fast lauter Kolonialengländer auf

dem Schiff. Bekanntschaft mit jüdischem Kaufmann

Haye aus Australien und einigen Amerikanern.

17, 18, 19. übel von schlechtem Futter. Hoher Seegang

und Regen 19. morgens Stromboli schön sichtbar.

Nachmittags 6 Uhr Neapel. Vesuv in grauen Wolken

Himmel bedeckt. Dabei so kalt und unfreundlich,

dass man froh ist, auf dem Schiff bleiben zu

können. Engländer aus Australien entpuppt

sich als Mecklenburger. Nachricht von Eisen-

bahnstreik in Frankreich und immer neuen

Repressalien an der Ruhr; wie mag es gehen?

In Marseille Toulon Leute freundlich, in Marseille

gefährlich deutsch zu sprechen. Bahnhofdirektor des

Güterverkehrs weigert sich, unser Gepäck nach Berlin,

ja selbst nach Zürich zu befördern.

22—28 Aufenthalt in Barcelona viel Mühe aber liebe

Menschen [Terradas, Campalang, Lana, Tirpitz, Tröltsch]

Volkslieder Tänze Refectorium. Schön war's!

27. Einstein with local children in L'Espluga de Francolí, Catalonia, Spain, 25 February 1923 (courtesy of the Spanish Ministry of Education, Culture and Sport. Archivo General de la Administración, Fondo Medios de Comunicación Social del Estado, signatura F-03198-010-00017 [Casemiro Lana Sarrate]).

colorful picture of small rectangular bath houses and larger houses along the beach. Radiant sunshine. Sense of liberation. Wife better, is being devotedly cared for by Mrs. Muschli.[261] Visit with governor (broad-faced Oriental) and a few consuls.

16th. Morning departure on "Ormuz" (Oriental line).[262] Bad grub. Almost exclusively English colonials on board, acquaintance with Jewish businessman Haye from Australia and a few Americans.

17th, 18th, 19th. Unwell from bad grub. High seas and rain. 19th in the morning, Stromboli nicely in view. Afternoon 6 p.m., Naples. Vesuvius in gray clouds, overcast sky. At the same time, so cold and unpleasant that one is happy to be able to remain on board the ship. Englishman from Australia turns out to be from Mecklenburg. News of railway strike in France and ever more reprisals along the Ruhr; how will things continue?[263] In ⟨Marseille⟩ Toulon people friendly; in Marseille, dangerous to speak German. Director of freight depot refuses to dispatch our luggage to Berlin, or even to Zurich.[264]

22nd–28th. Sojourn in Barcelona. Very tiring, but dear people. Terradas, Campalans, Lana, Tirpitz's daughter. Folk songs, dances. Refectory.[265] How nice it was![266]

## 40

1. März  Ankunft in Madrid. Abfahrt von Barcelona

herzlicher Abschied, Terradas, deutscher Konsul mit Trypsty' Gatten etc

3. März. Erster Vortrag in Universität.

1st March. Arrival in Madrid. Departure from Barcelona, cordial farewells. Terradas, German consul with Tirpitz's daughter, etc.[267]

3rd March. First lecture at university.[268]

28. Reception for Einstein at the Spanish Royal Academy, Madrid, 4 March 1923. Front row includes King Alfonso XIII, Einstein, José Rodríguez Carracido, and Blas Cabrera (courtesy of Granger Collection).

4. März Autofahrt mit Kocherthalers. Antwort
auf Cabreras Akademierede geschrieben. Nachmittags Akademiesitzung mit Königs Vorsitz.
Wunderbare Rede des Akad. Präsidenten. Dann
Thee bei ärztl. Gesellschaftsdame. Abends gehause
aber ganz katholisch.

5. Vormittags mathematische Gesellschaft
Ehrenmitglied. Diskussion über allgemeine
Relativität. Essen bei Kuno. Besuch bei Kochal
Wunderbarer alter Kopf. Schwer krank. Vortrag.
Abends Essen eingeladen von Koc, Vogel. Gutherziger
humorvoller Pessimist.

6. Durch viele Lügen eachierter Ausflug
nach Toledo. Einer der schönsten Tage meines
Lebens. Strahlender Himmel. Toledo wie ein
Märchen. Ein begeisterter alter Mann, der Bedeutendes
über den Greco geschrieben haben soll führt uns
Strassen und Marktplätze, Blicke auf die Stadt, Tago
mit Steinbrücken, steinbedeckte Hügel, liebliche
Ebene Dom, Synagoge, Sonnenuntergang auf der Heimfahrt mit glühenden Farben. Gärtchen mit Aussicht
bei Synagoge. Herrliches Bild von Greco in kleiner
Kirche (Beerdigung eines Adels) gehört zum tiefsten,
was ich je sah. Wundervoller Tag.

4th March. Car ride with Kocherthalers. Reply to Cabreras. Wrote speech for academy. In the afternoon, academy session chaired by King. Wonderful speech by the acad[emy] president.[269] Then tea at arist[ocratic] lady companion's.[270] Evening at home but very Catholic.[271]

5th. Morning. Mathematical society honorary membership. Discussion about general relativity. Meal at Kuno's. Visit with Kuchal. Wonderful old thinker. Seriously ill. Lecture. In the evening, dinner invitation by Mr. Vogel.[272] Kind-hearted, humorous pessimist.

6th. Excursion to Toledo concealed through many lies. One of the finest days of my life. Radiant skies. Toledo like a fairy tale. An enthusiastic old man, who had supposedly written something of import about ⟨Goy⟩ El Greco, guides us. Streets and market place, views onto the city, the Tagus with stone

29. Invitation to audience with Maria Christina of Austria, Queen Mother of Spain, Madrid, 6 March 1923 (with permission of the Albert Einstein Archives, the Hebrew University of Jerusalem).

bridges, stone-covered hills, charming plain, cathedral, synagogue, sunset on the homeward trip with glowing colors. Small garden with vistas near synagogue. Magnificent painting by El Greco in small church (burial of a nobleman) is among the profoundest images I have ever seen.[273] Wonderful day.

7. 12 Uhr Audienz bei König und Königin Mutter. Letztere zeigt ihre Wissenschaft. Man merkt dass niemand ihr sagt, was er denkt. Der König einfach und würdig, ich bewunderte ihn in seiner Art. Nachmittags dritter Universitätsvortrag, andächtiges Publikum, das sicher fast nichts verstehen konnte, weil die letzten Probleme behandelt wurden. Abends grosser Empfang beim deutschen Gesandten. Gesandter u. Familie prächtige schlichte Leute. Gesellschaft Strafe wie immer.

8. Ehrendoktor. Acht spanische Reden mit zugehörigem bengalischem Feuer. Lange Rede des d. Gesandten aber schliesslich gute über deutsch-span. Beziehungen, über ... nicht deutsch. Nichts historisches. Abends zum Besuch bei techn. Studenten. Reden und nichts als Reden, aber gut gemeint. Abends Vortrag. Dann bei Kunst musizieren. Ein Künstler (Direktor des Konservatoriums) Porus spielte herrlich Violine.

9. Ausflug nach Gebirge und Eskorial. Herrlicher Tag. Abends Empfang im Studentenheim mit Reden von Ortega. und mir.

10. Prado (Bilder von Velasquez und Greco hauptsächlich betrachtet) Abschiedsbesuche. Essen bein deutschen

30. Diploma of honorary doctorate from the University of Madrid, 8 March 1923 (with permission of the Albert Einstein Archives, the Hebrew University of Jerusalem).

7th. 12 p.m. audience with the King and Queen Mother. The latter demonstrates her science. One notices that no one tells her what they think. The King, simple and dignified, I admire him for his manner. In the afternoon, third university lecture; devout audience who surely couldn't understand virtually anything, because the latest problems were discussed. In the evening, grand reception at the German ambassador's. Ambassador & family magnificent, modest people.[274] Socializing as punishing as always.

8th. Honorary doctorate. Truly Spanish speeches with associated Bengali fire. Long speech but good in respect to content by the ambassador about German-Spanish relations; truly German. Nothing rhetorical. Then, visit with the techn[ology] students. Speeches and nothing but speeches, but well intentioned. In the evening, lecture. Then music making at Kuno's. A professional (director of the conservatory) Poras played violin exquisitely.[275]

9th. Excursion to mountains and Escorial. Glorious day. In the evening, reception at the student residence with speeches by Ortega and me.[276]

10th. Prado (mainly viewed paintings by Vélasquez and El Greco). Farewell visits. Meal at the German

Gesandten. Abends mit Lina und Uhlmanns in
primitivem *kleinen* Tanzlokal. Fröhlicher Abend.

11. Prado (Herrliche Werke von Goya, Raphael, Fra
Angelico)

12. Abreise nach Zaragoza.

ambassador's. Evening with Lina and the Ullmanns in primitive little dance hall.[277] Merry evening.

11th. Prado (splendid masterpieces by Goya, Raphael, Fra Angelico).[278]

12th. Trip to Zaragoza.[279]

31. Einstein with faculty of the University of Madrid, 8 March 1923. Front row: Miguel Vegas, José Rodríguez Carracido, Einstein, Luis Octavio de Toledo, and Blas Cabrera. Back row: Edmundo Lozano Rey, Josep M. Plans, José Madrid Moreno, E. Lozano Ponce de León, Ignazio González Martin, Julio Palacios, and Angel del Campo (with permission of the Albert Einstein Archives, the Hebrew University of Jerusalem).

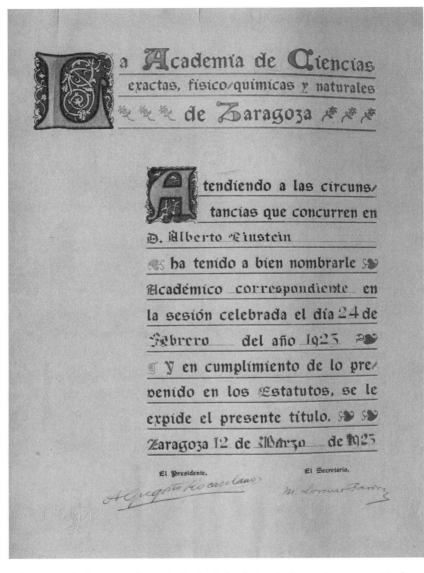

32. Diploma of the Academy of Exact, Physicochemical and Natural Sciences of Zaragoza, 12 March 1923 (with permission of the Albert Einstein Archives, the Hebrew University of Jerusalem).

# Additional Texts

TEXT 1. From Sanehiko Yamamoto[1]

[Tokyo, 15 January 1922]

The negotiations.

Kaizosha has the honor of most courteously inviting Professor Dr. Albert Einstein to Japan and asking him to give some lectures. We are both agreed about definitely satisfying the following commitment:

1. There are planned
   a) One scientific lecture, in Tokyo during a period of six days, in particular, about three hours each day; and
   b) Six popular talks, once each in Tokyo, Kyoto, Osaka, Fukuoka, Sendai, and Sapporo (for around two and a half hours).
2. If no inevitable obstacles arise, the speaker would embark on the trip at the end of ⟨August⟩ September or the beginning of ⟨September⟩ October[2] 1922. The sojourn in Japan will last about one month.
3. The honorarium (including the voyage as well as accommodation costs) amounts to two thousand (£2,000) English pounds.

Kaizosha transfers to the speaker half of the total together with this certificate through the "Yokohama Specie Bank" in London. The remainder will be presented to him immediately upon his arrival in Japan.

If it is impossible to come to Japan owing to unavoidable difficulties, the advance in the amount of English £1,000 should be repaid to *Kaizosha*.

| In great respect, we both undersign: | S. Yamamoto |
|---|---|
| Berlin, the — 1922 | (The representative |
| Tokyo, the 15th of Jan. 1922. | of Kaizosha.) |

TEXT 2. Report on a conversation with Prof. Einstein on the
day of his departure to Japan, on 29 September 1922[3]

[Berlin,] 12 October 1922

Prof. *Einstein* declares that he is prepared to accept the invitation of
Dr. Arthur *Ruppin* to visit Palestine.[4] It will be possible for him to
arrange his itinerary in such a manner that, on the return trip from
Batavia,[5] he will be able to stay in Palestine for 10 days . . . Einstein
would like to stress that this brief stay should not be confused with
his actual trip to Palestine.[6] "One should only travel to Palestine directly and not on the occasion of a trip to other countries;" furthermore, he knows that in 10 days he cannot form an opinion on the
issues which really interest him. . . . Moreover, it seems necessary to
me, that our opinion on the establishment of the university[7] be imparted to him as an official view. He will then campaign for this cause
and be prevented from making random opinions on the plans for the
university his own.

Kurt Blumenfeld

TEXT 3. Speech at Reception in Singapore[8]

[Singapore, 2 November 1922]

I feel highly pleased at the hearty welcome you, Mr. Manasseh Meyer,
and all here, in your mansion, have given to me and to Mrs. Einstein.
I have also been greatly touched by the impressive words you have
made use of in the address which you have presented to me.[9] On behalf of Mrs. Einstein and myself, I beg to thank you for the kind reception you have given to us. I am agreeably surprised to find here in
the Far East such a happy unity amongst our brethren.[10] Concerning
your personal remarks addressed to me, they made me feel all the
more happy on account of their recognition of intellectual ambitions,
which is one of the finest traditions of our race. (Hear, hear.)

The flattering terms which you have used regarding my theory should not really have been addressed to me but to all the scientists of the last century and which was the result of the progress of science through the ages. I am glad to say that science is the property of all nations, and is not in any way endangered by international strife, for it always has a healing influence on those people who look beyond the horizon. If science is preeminent through its universal predomination, then one may ask "Why do we need a Jewish University?"[11] Science is international but its success is based upon institutions which are owned by nations. If, therefore, we wish to promote culture, we have to combine and to organise institutions with our own power and means. We need to do this all the more on account of the present political developments and especially in view of the fact that a large percentage of our own sons are refused admission into the Universities of other nations.[12] (Shame.) Up to now, as individuals, we have helped as much as possible in the interests of culture, and it would be only fair to ourselves if we, now, as a race, add to culture through the medium of our own institution. (Applause.) With this end in view let us work together with all those prominent men who are already devoting all their energy to the fulfillment of this great ideal. Once more I heartily thank you all for the great esteem which you have shown to me.

TEXT 4. "Chat about My Impressions in Japan"[13]

Manuscript completed on or after 7 December 1922.
Published January 1923[14]

I have been traveling much around the world in the last few years, actually more than befits a scholar. The likes of me really ought to stay quietly put in his study and ponder. For my earlier trips there was always an excuse that could easily pacify my not very susceptible conscience. But when Yamamoto's invitation to Japan arrived,[15]

I immediately resolved to embark on such a great voyage that must demand months, even though I am unable to offer any excuse other than that I would never have been able to forgive myself for letting a chance to see Japan with my own eyes pass unheeded.

Never in my life have I been more envied in Berlin, and genuinely so, than the moment it became known that I was invited to Japan. For in our country this land is shrouded more than any other in a veil of mystery. Among us we see many Japanese, living a lonely existence, studying diligently, smiling in a friendly manner. No one can fathom the feelings concealed behind this guarded smile. And yet it is known that behind it lies a soul different from ours that reveals itself in the Japanese style, as is manifest in numerous small Japanese products and Japanese-influenced literature becoming fashionable from time to time. All the things I knew about Japan could not give me a clear picture. My curiosity was in utmost suspense when, on board the *Kitanu Maru*, I passed through the Japanese straits and saw the countless dainty green islets glowing in the morning sun. But glowing most of all were the faces of all the Japanese passengers and the ship's entire crew. Many a delicate little woman, who otherwise would never be seen before breakfast time, was roaming restlessly and blissfully about on deck at six o'clock in the morning, heedless of the harsh morning wind, in order to catch the first possible glimpse of her native soil. I was moved to see how overcome they all were with deep emotion. A Japanese loves his country and his nation above all else; and despite his linguistic proficiency and great curiosity about everything foreign, away from home he still does feel more alien than anyone else. How is this explained?

I have been in Japan for only two weeks now,[16] and yet so much is still as mysterious to me as on the first day. Some things I have learned to understand, though, most of all the shyness that a Japanese feels in the company of Europeans and Americans. At home our entire upbringing is directed toward our being able to tackle life's struggles successfully as individuals under the best possible circumstances. Particularly in the cities—individualism in the extreme, cutthroat competition exerting one's utmost energy, feverish laboring to acquire as much luxury and indulgences as possible. Family bonds are loosened, the influence of artistic and moral traditions on daily life is relatively slight. The isolation of the individual is seen as a necessary consequence of the struggle for survival; it robs a person of that carefree happiness that only absorption in a community can offer. The predominantly rationalistic education— indispensable for practical living under our circumstances—lends this attitude of the individual even more poignancy; it makes one even more aware of the isolation of the individual.

Quite the contrary in Japan. The individual is left far less to his own devices than in Europe or America. Family ties are very much closer than at home, even though they are actually only provided very weak legal protection. But the power of public opinion is much stronger here than at home, ensuring that the family fabric is not undone. Public and private reputation helps force to completion what is mostly already sufficiently secured by a Japanese upbringing and an innate kindheartedness.

The cohesion of extended families in material respects, mutual support, is facilitated by an unpretentiousness of the individual in regard to housing and food. A European can generally accommodate one person in his apartment without there being perceptible disruption to the household order. Thus a European man can at best mostly only care for his wife and children. Often wives, even women of higher status, must help earn a living and leave the children's education to the servants. It is rare that adult siblings, let alone more distant relatives, provide for one another.

But there is a second reason, as well, that makes closer protective ties between individuals easier in this country than where we live. It is the idiosyncratic Japanese tradition of not expressing one's feelings and emotions but staying calm and even-tempered in all circumstances. This is the basis upon which many persons, even those not in emotional harmony with one another, can live under a single roof without embarrassing frictions and conflicts arising. Herein lies, it appears to me, the deeper meaning of the Japanese smile, which is so mysterious to a European.

Does this upbringing to suppress the expression of an individual's feelings lead to an inner impoverishment, a suppression of the individual himself? I do not think so. The development of this tradition was surely facilitated by a refined sensitivity characteristic of this nation and by an intense sense of compassion, which seems to be more vivid than for Europeans. A rough word does not injure a European any less than it does a Japanese. The former immediately counters by going on the offensive, amply repaying in kind. A Japanese withdraws, wounded, and—weeps. How often is the Japanese inability to utter sharp words interpreted as falseness and dishonesty!

For a foreigner like me it is not easy to delve deeply into the Japanese mind. Being received everywhere with the greatest attention in festive garb, I hear more carefully weighed words than meaningful ones that inadvertently slip out of the depths of the soul. But what escapes me in direct experience with people is completed by the impressions of art, which is so richly and diversely appreciated in Japan as in no other country. By "art" I mean all things of permanence that human hands have created here whether by aesthetical intent or secondary motivation.

In this regard I hardly cease to be astonished and amazed. Nature and people seem to have united to bring forth a uniformity in style as nowhere else. Everything that truly originates from this country is delicate and joyful, not abstractly metaphysical but always quite closely connected with what is available in nature. Delicate is the landscape with its small green islets or hills, delicate are the trees, delicate the most carefully farmed land with its precise little parcels, but most especially so the little houses standing on the land; and, finally, the people themselves in their speech, their movements, their attire, including all objects of which they make use. I took a particular liking to the Japanese house with its very segmented smooth walls, its many little rooms laid out throughout with soft mats. Each little detail has its own sense and import there. Add to that the dainty people with their picturesque smiling, bowing, sitting—all things that can only be admired but not imitated. You, oh foreigner, try to do so in vain!—Yet Japan's dainty dishes are indigestible to you; better content yourself with watching.—Compared to our people, the Japanese are cheerful and carefree in their mutual relations—they live not in the future but in the present. This cheerfulness always expresses itself in refined form, never boisterously. Japanese wit is directly comprehensible to us. They too have much sense for the droll, for humor. I was astonished to note that, regarding these psychologically surely deeply lying things, there is no great difference between the Japanese and the European. The soft-heartedness of the Japanese reveals itself here in that his humor does not assume a sarcastic note.

Of greatest interest to me was Japanese music, which developed partly or entirely independently of ours. Only upon listening to completely strange art does one come closer to the ideal of separating the conventional from the essential conditioned by human nature. The differences between Japanese music and ours are indeed fundamental. Whereas in our European music, chords and architectonic structure appear to be universal and indispensable, they are absent in Japanese music. On the other hand, they have in common the same thirteen tonal steps into which the octaves are divided. Japanese music seems to me to be a kind of emotive painting of inconceivably direct impressions. The exact tonal pitch does not even seem to be so absolutely decisive for the artistic effect. It much rather suggests to me a stylized depiction of the passions expressed by the human voice as well as by such natural sounds that conjure up in the human mind emotive impressions, such as birdsong and the breaking of waves in the sea. This impression is amplified by the important role played by percussive instruments, which have no specific tonal pitch of their own but are particularly suited for rhythmic characterization. The main attraction to me of Japanese music lies in the extremely refined rhythms. I am fully aware of the circumstance that the most intimate subtleties of this kind of music still elude me. Added to the fact that long experience is always presupposed for distinguishing the purely conventional from a performer's personal expression, the relation to the spoken and sung word, which plays a considerable role in most Japanese pieces of music, also escapes me. As I see it, a characteristic of the artistic approach of the Japanese soul is the unique appearance of the mellow flute, not the much more strident wind instruments made out of metal. Here, too, is manifested the particular characteristic preference for the mellifluous and the dainty that is especially prominent in Japanese painting and design of products for everyday life. I was most greatly affected by the music when it served as accompaniment to a theater piece or a mime (dance), particularly in a Noh play. What stands in the way of the development of Japanese music into a major form of high art is, in my opinion, a lack of formal order and architectonic structure.

To me, the area of most magnificence in Japanese art is painting and wood carving. Here it is truly revealed that the Japanese is a visual person who delights in form, who untiringly refashions events in artistic form, converted into stylized lines. Copying nature in the sense of our realism is foreign to the Japanese, just as is religious repudiation of the sensual, despite the influence of Buddhism from the Asian continent, which is intrinsically so foreign to the Japanese soul. For him, everything is experienced in form and color, true to Nature and yet foreign to Nature insofar as stylization broadly predominates. He loves clarity and the simple line above all else. A painting is strongly perceived as an integral whole.

I have only been able to mention a few of the great impressions I had during these weeks, saying nothing about political and social problems. On the exquisiteness of the Japanese woman, this flower-like creature—I have also remained reticent; for here the common mortal must cede the word to the poet. But there is one more thing that weighs on me. The Japanese rightfully admires the intellectual achievements of the West and immerses himself successfully and with great idealism in the sciences. But let him not thereby forget to keep pure the great attributes in which he is superior to the West—the artful shaping of life, modesty and unpretentiousness in his personal needs, and the purity and calm of the Japanese soul.

TEXT 5. To Sanehiko Yamamoto[17]

Kyoto, 12 December 1922

Esteemed Mr. Yamamoto,

Under the impression of the great services you have rendered to me and my wife during our presence in Japan, I consider it an absolute duty to inform you of the following. The ship[18] is departing from Moji in only 16 days' time and in the interim I have nothing for you to accomplish on our behalf. So I feel it is unfair of me to have us impose upon Mr. Inagaki and his wife during this interval.[19] Much as I love them both, I ask you please, in order to relieve my conscience, to leave my wife and me *alone* in Kyoto during this quiet period. You are truly doing enough for us by making possible for us such a long stay in this wonderful city. I also ask you please not to have anyone travel to Fukuoka and Moji *purely as a personal favor to us*.

At this time I would like to express my profound gratitude to you for providing us with the opportunity to see this wonderful country and for making our lives more pleasant throughout in such a generous and caring manner.

Cordial greetings to you, yours,
A. Einstein.

My wife was not in Osaka yesterday, because I had ordered her to remain in Kyoto. I did this because it was not known to me in time that an official function was supposed to take place in Osaka.[20] My wife was very unhappy about having been the source of a disruption through no fault of her own.[21]

TEXT 6. To Hans Albert and Eduard Einstein[22]

Kyoto, 17 December 1922

Dear Children,

Now you, d[ear] Albert, have been a student for a couple of months already.[23] I often think of that with pride. The voyage is wonderful, even though Japan is quite exhausting. I have already given 13 lectures. I am very glad that I left you, d[ear] Albert, in Zurich; as I would not have been able to pay much attention to you anyway, and your studies mean more to you than any trip, no matter how nice, in which you'd have had to make official appearances in so many instances.[24] The Japanese do appeal to me, by the way, better than all the peoples I've met up to now: quiet, modest, intelligent, appreciative of art, and considerate, nothing is for the sake of appearances, but rather everything is for the sake of substance. So now you really will be getting the Nobel Prize.[25] Start looking for a house.[26] The rest will be invested somewhere in your names. Then you'll be so rich that, God knows, I may have to hit you up some day, depending on how things go. After my return home (end of March or beginning of April), I will soon have to travel to Stockholm to accept the prize. When I travel to Geneva then,[27] I shall visit you, of course; I'm already looking forward to that. Then we can also consult about what we'll be doing next summer. I am determined not to gallivant around the world so much anymore; but am I going to be able to pull that off, too?

You rascals didn't write me at all; now it's too late for Asia. If you want to write me before my return home to Germany, e.g., about the house, then write to Spain (University of Madrid) or—if you want to write fast—to the Zionist Organisation in Jerusalem. I am enclosing for you, d[ear] Tete,[28] a few postage stamps collected along the way.

Affectionate regards to you and Mama[29] from your

⟨Albert⟩ Papa.

TEXT 7. To Wilhelm Solf[30]

[Miyajima, 20 December 1922][31]

[ . . . ]

I hasten to dispatch to you the more detailed information as a supplement to my reply by telegraph.[32] Harden's statement is certainly awkward for me, in that it aggravates my situation in Germany; nor is it completely correct, but neither is it completely wrong.[33] Because people who know the situation in Germany well are indeed of the opinion that a certain threat to my life does exist. Admittedly, I did not assess the situation the same way prior to Rathenau's murder as I did afterwards. A yearning for the Far East led me, in large part, to accept the invitation to Japan; another part was the need to get away for a while from the tense atmosphere in our homeland for a period of time, which so often places me in difficult situations. But after the murder of Rathenau, I was certainly very relieved to have an opportunity for a long absence from Germany, taking me away from the temporarily heightened danger without my having to do anything that could have been unpleasant for my German friends and colleagues.

[ . . . ]

TEXT 8. To Jun Ishiwara[34]

Moji, [between 23 and 29 December 1922][35]

To my dear colleague Ishiwara as a memento, with whom I saw so many beautiful things, collaborated, and chatted many pleasant hours away. He is one of those few with whom I would very much like to ponder and work in a comradely spirit; because, in spite of all the differences in origin and tradition, a mysterious harmony exists between us.

Albert Einstein
Moji 1922.

## TEXT 9. To Bansui Tsuchii (Doi)[36]

[On board S.S. *Haruna Maru*,] 30 December 1922

Very esteemed Mr. Tsuchii,

With great joy and admiration did I read the translation of your thoughtful poem[37] and your very friendly letter. It does not matter that you hugely overrate my accomplishments,[38] if the words simply emanate from a pure soul. The scientific quest really is different from that of an artist. The latter will evolve with certainty if he has the ability to see and feel, the power to create, and the stamina and love of perfect creativity. Science, however, is like riddle-guessing or even playing the lottery. It is a rare experience of happiness when one finds something of real value. Many a highly talented young man labors until ripe old age without the severe goddess unveiling anything of her deep secrets to him; she is unpredictable and inquires little about the merit earned from a devoted search for truth. And the little she has entrusted to me must appear gigantically magnified to the uninformed, who do not know about the achievements of the forerunners and fellow seekers. Be that as it may—I am delighted about your enthusiastic words.

I found exceptionally fine what you said about your beautiful land and the peculiar transitional state in which it currently finds itself. But I do believe you were a little too harsh in your characterization. In nurturing Western sciences over the course of these couple of decades, Japan has already ascended to a high level, and the profoundest of problems are being tackled. Japan is not merely borrowing the external elements of civilization from the West![39] An inundation of foreign culture is dangerous for any country in which its own high values are too easily underrated and forgotten—I am referring to the artistic, social, and ethical traditions of your country that I admire and cherish so much. In such matters the Japanese does not recognize his superiority over the European; it would be of great service to bring this to his consciousness so that he can feel that, with the indiscriminate adoption of European customs, great values are being jeopardized. Japan may acquiesce to the civilizing spirit of Europe and America, but it must know that its soul is far more valuable than these external shimmering trifles.

With joy and with a trembling hand I accept the splendid reproductions of Japanese-Chinese works of art which you have bestowed upon me.[40] They will accompany me on this voyage and they will soften the harshness of the transition to Europe. The Japanese artistic hand is of incomparable delicacy.

My extant collected works will soon appear in Japanese,[41] and I will gladly permit myself to have them sent to you, although it will be difficult to insert a dedication. But I will try to do so. I am attaching a small card for your son.[42]

> Please be greeted from the bottom of my heart and
> please accept my warmest thanks, from your,
> A. Einstein

TEXT 10. To Eiichi Tsuchii (Doi)[43]

[On board S.S. *Haruna Maru*, 30 December 1922][44]

He who is familiar with the exertions of pondering scientific problems never feels empty and alone; he also gains a firm foothold against the vicissitudes of fate.

> A greeting for the young man E-i-ichi,
> Albert Einstein

TEXT II. To Yoshi Yamamoto[45]

[On board S.S. *Haruna Maru*, 30] December 1922[46]

You, esteemed Mrs. Yamamoto, will always epitomize for me the ideal form of Japanese femininity. Quiet, cheerful, and flowerlike, you are the soul of your home, which looks like a jewelry case, within which, like jewels, lie your darling little children.[47] In you I rightly see the soul of your people and the embodiment of its ancient culture, directed primarily toward grace and beauty.

Yours,
Albert Einstein.

TEXT I2. Speech at Jewish Reception in Shanghai[48]

[Shanghai, 1 January 1923]

I was asked to say a few words about the University at Jerusalem.[49] I arrived at the conviction that such an institution was necessary from personal experience. While I was studying in Switzerland, I did not even know that I was a Jew.[50] I was satisfied to know only that I was a man. When I came later to Berlin, I became aware that many like myself felt the spiritual need.[51] They required something whereby to make their Jewish consciousness articulate and audible, and as the need could not be supplied, some attempted to stifle it by artificial means, which brought them no satisfaction.

Then came Zionism and brought a new harmony into the souls of many.[52] Now the Jewish University will provide the focus for the Jewish spirit and will enable Jewish scholars to find their bearings. It will not be so much a school for students as a rallying point for Jewish scholarship and an authoritative center of Jewish thought that will help to define and clarify our outlook throughout the wide

33. Nobel Prize in Physics, medal, 10 December 1922 (with permission of the Albert Einstein Archives, the Hebrew University of Jerusalem).

world. And from it, influences will radiate that will enliven and inspirit the diverse communities of scattered Israel.

TEXT 13. To Svante Arrhenius[53]

Near Singapore, 10 January 1923

Esteemed Colleague,

The news of the award of the Nobel Prize reached me via telegraph on the *Kitano Maru* shortly before my arrival in Japan.[54] I am very pleased—among other reasons, because the reproachful question: why don't you get the Nobel Prize? can no longer be posed to me. (I would reply each time: Because *I* am not the one who awards the prize.)

Mrs. Hamburger[55] informed me that you were so kind as to invest the money temporarily.[56] Thank you very much for this kind solicitude. You (and Bohr)[57] also wrote that the official award ceremony has been scheduled for June, which I appreciate very much. I am returning from this wonderful trip at the latest at the beginning of April. I am very enthralled by the land and people of Japan, everything is so subtle and strange. And how conducive to thinking and working the long sea voyage is—a paradisiacal state without correspondence, visits, meetings, and other inventions of the devil! It is a special pleasure for me that I receive the prize together with my admired and beloved Bohr.

In the happy prospect of a cheerful reunion, at the latest in Stockholm, I am, with all due respect and kind regards, yours,
A. Einstein.

TEXT 14. To Niels Bohr[58]

Near Singapore, 10 January 1923

Dear, or rather, beloved Bohr!

Your affectionate letter[59] reached me shortly before my departure from Japan.[60] I can say without exaggeration that it pleased me as much as the Nobel Prize. I find your fear of possibly getting the prize before me especially endearing—that is genuinely Bohr-like. Your new studies on the atom[61] accompanied me on my journey, and my love for your mind has grown even more. I now believe I finally understand the connection between electricity and gravitation.[62] Eddington came closer to the gist of it than Weyl.[63]

This is a splendid voyage. I am delighted by Japan and the Japanese and am sure that you would be, too. What's more, such a sea voyage is a splendid existence for a ponderer—it's like a monastery. Add to that the cajoling warmth near the equator. Warm water drips languidly down from the heavens and spreads calm and vegetative dozing—this little letter is a proof thereof.

Warm regards. To a happy reunion, in Stockholm at the latest. Your admiring
A. Einstein.

TEXT 15. To Nippon Puroretaria Domei[64]

[On board S.S. *Haruna Maru*, 22 January 1923][65]

Dear friends,

I was unable to answer your letter[66] earlier because I had lost both it and the address. Mr. Yamamoto[67] has again furnished me with the address, so I would like to reply to your questions, but am unable to recall them in detail.

First, I must point out that my observations about Japanese social and political conditions are so very limited that even I cannot rely on my own judgment. Concerning the first point I observed two matters which at a glance seem to be incompatible. There is neither conspicuous poverty nor a lack of money, but nevertheless piecework carried out at home is dreadfully ill-paid for the most part. As far as my observations showed, I believe that this riddle can be interpreted as the result of the fact that the people have few desires, their way of life is suitable, and in addition they are particularly moderate in their consumption of alcohol. Be that as it may, in any case I believe that this country will become increasingly industrialized, and owing to the political situation, it will become necessary to organize the working classes. If this organization is to be of value to the whole nation, it must not turn into a malicious movement which is carrying out opposition merely for the sake of opposition, such as happened with us in Europe for a long time. You must realize in particular that the main factor behind the lowering in wages for home work lies in the overpopulation of this country, and therefore it cannot be done away with by purely political methods. On the other hand, the struggle against militarism seems to me to be a purely political issue. In my opinion this constitutes a real danger to this nation. This is because, owing to its geographical position, Japan is fortunate enough to require little military protection. The Washington Conference created the first opportunity enabling us to hold out some hope on this matter.[68] I

**RED STAR LINE**
NEW TRIPLE SCREW STEAMER BELGENLAND 26.500 TONS.

34. Postcard to Arthur Ruppin, Jerusalem, 3 or 5 February 1923. In Einstein's sketch, he added "halo" next to the drawing of himself and "Mrs. Ruppin" next to the drawing of Hannah Ruppin (with permission of the Albert Einstein Archives, the Hebrew University of Jerusalem).

am convinced that in the future efforts by the people will be linked to international cooperation as well as organization, and will never combine with military planning. I hope that Japan will draw a conclusion from this for its own sake and for the sake of all the countries in the world.

Special regards,
A. Einstein

### TEXT 16. To Arthur Ruppin[69]

[Jerusalem, 3 or 5 February 1923][70]

Dear Mr. Ruppin,

We are spending beautiful, unforgettable days in Palestine with the sun shining brightly and in joyous company. Your wife[71] is standing next to me and is peeking at what I am writing about her. She is counting the days until your return.[72] Yours,

A. Einstein.

### TEXT 17. "Prof. Einstein on His Impressions of Palestine"[73]

[Berlin, 24 April 1923]

My Impressions of Palestine[74]
By Albert Einstein

I cannot begin these notes without expressing my heartfelt gratitude to those who have shown so much friendship toward me during my stay in Palestine. I do not think I shall ever forget the sincerity and warmth of my reception—for they were to me an indication of the harmony and healthiness which reigns in the Jewish life of Palestine.

No one who has come into contact with the Jews of Palestine can fail to be inspired by their extraordinary will to work and their determination, which no obstacle can withstand. Before that strength and spirit there can be no question of the success of the colonization work.

The Jews of Palestine fall into two classes—the urban workers and the village colonizers. Among the achievements of the former, the city of Tel Aviv made a singularly profound impression on me.[75] The rapidity and energy which has marked the growth of this town has been so remarkable that Jews refer to it with affectionate irony as "Our Chicago."

A remarkable tribute to the real power of Palestine is the fact that those Jewish elements which have been resident in the country for decades stand distinctly higher, both in the matter of culture and in their display of energy, than those elements which have only recently arrived.

And among the Jewish "sights" of Palestine none struck me more pleasantly than did the school of arts and crafts, Bezalel, and the Jewish workingmen's groups.[76] It was amazing to see the work that had been accomplished by young workers who, when they entered the country, could have been classified accurately as "unskilled labor." I noted that beside wood, other building material is being produced in the country. But my pleasure was tempered somewhat when I learned of the fact that the American Jews who lend money for building purposes exact a high rate of interest.[77]

To me there was something wonderful in the spirit of self-sacrifice displayed by our workers on the land. One who has actually seen these men at work must bow before their unbreakable will and before the determination which they show in the face of their difficulties—from debts to malaria.[78] In comparison with these two evils, the Arab question becomes as nothing. And in regard to the last I must remark that I have myself seen more than once insurance of friendly relations between Jewish and Arab workers. I believe that most of the difficulty comes from the intellectuals—and, at that, not from the Arab intellectuals alone.[79]

The story of the struggle against malaria constitutes a chapter by itself. This is an evil which affects not only the rural, but also the urban population. During my visit to Spain some time ago, we submitted to the Spanish Jews a proposition that they send, at their expense, a specialist on the subject of malaria to Palestine, and that this specialist should carry on his work in connection with the work of the University in Jerusalem.[80] The malaria evil is still so strong that one may say that it weakens our colonization work in Palestine by something like a third.

But the debt question is particularly depressing. Take for instance the workers of the colony of Deganiah.[81] These splendid people groan under the weight of their debts, and must live in the direst need in order not to contract new ones. One man, even with moderate means, could, if he were largehearted enough, relieve this group of its heartbreaking burden. The spirit which reigns among the land and building workers is admirable. They take boundless pride in their work and have a feeling of profound love for the country and for the little locality in which they work.

In the matter of architectural taste, as displayed in the buildings, in the towns and on the land, there has been not a little to regret. But in this regard the engineer, Kaufman,[82] has done a great deal to bring good taste and a love of beauty into the buildings of Palestine.

To the government[83] considerable credit must be accorded for its constructions of roads and paths, for its fight against malaria and, in general, for its sanitary work as a whole. Here the government has no light task before it. One can hardly find another country which, being so small, is so complicated by virtue of the divisions among its own population as well as by virtue of the interest taken in it by the outside world.

The greatest need of Palestine today is for skilled labor. No academic forces are needed now. It is hoped that the completion of the technikum[84] will do a great deal toward meeting the need of the country for trained workmen.

I am convinced that the work in Palestine will succeed in the sense that we shall create in that country a unified community which shall be a moral and spiritual center for the Jewries of the world. Here, and not in its economic achievement, lies, in my opinion, the significance of this work. Naturally we cannot neglect the question of our economic position in Palestine, but we must at no time forget that all this is but a means to an end. To me it seems of secondary importance that Palestine shall become economically independent with the greatest possible speed. I believe that it is of infinitely greater importance that Palestine shall become a powerful moral and spiritual center for the whole of the Jewish people. In this direction the rebirth of the Hebrew language must be regarded a splendid achievement. Now must follow institutions for the development of art and science. From this point of view we must regard as of primary importance the founding of the University which, thanks largely to the enthusiastic devotion of the Jewish doctors of America, can begin its work in Jerusalem.[85] The University already possesses a journal of science which is produced with the earnest collaboration of Jewish scientists in many fields and in many countries.[86]

Palestine will not solve the Jewish problem, but the revival of Palestine will mean the liberation and the revival of the soul of the Jewish people. I count it among my treasured experiences that I should have been able to see the country during this period of rebirth and reinspiration.

# Chronology of Trip

| November 17 | Arrives in Kobe. Train journey to Kyoto. |
|---|---|
| November 18 | Sightseeing tour of Kyoto. Train journey to Tokyo. |
| November 19 | Delivers first public lecture at Keio University. |
| November 20 | Reception in Einstein's honor held by the Imperial Academy of Sciences at the Koishikawa Botanical Gardens. |
| November 21 | Attends chrysanthemum-viewing garden party in the gardens of the Akasaka Imperial Palace. Greeted by Empress Teimei. |
| November 22 | Reception at Kaizo-Sha Publishing House. |
| November 24 | Delivers second public lecture at the Youth Assembly Hall in Tokyo. |
| November 25 | Delivers first scientific lecture at the Physics Institute of Tokyo Imperial University. |
| November 27 | Delivers second scientific lecture at the Physics Institute of Tokyo Imperial University. |
| November 28 | Reception at the Tokyo University of Commerce. |
| | Delivers third scientific lecture at the Physics Institute of Tokyo Imperial University. |
| November 29 | Attends tea ceremony. |
| | Reception at Waseda University in Tokyo. |
| | Delivers fourth scientific lecture at the Physics Institute of Tokyo Imperial University. |
| | Reception at the Tokyo Women's Higher Normal School. |

| November 30 | Visits the Division of Imperial Court Music at the Imperial Household Agency. |
|---|---|
| | Delivers fifth scientific lecture at the Physics Institute of Tokyo Imperial University. |
| | Reception held by students of all Tokyo universities. |
| December 1 | Delivers sixth scientific lecture at the Physics Institute of Tokyo Imperial University. |
| December 2 | Visits the Tokyo School of Technology. |
| | Train journey to Sendai. |
| December 3 | Delivers third public lecture at Sendai Civic Auditorium. |
| | Inscribes his signature with Japanese calligraphy brushes on the wall of a conference room at Tohoku Imperial University in Sendai. |
| | Train journey to the Matsushima Islands. |
| December 4 | Train journey to Nikko. |
| December 5 | Sightseeing at Lake Chuzenji. |
| December 6 | Sightseeing in Nikko. Return train journey to Tokyo. |
| December 7 | Train journey to Nagoya. |
| December 8 | Delivers fourth public lecture at the Gymnasium for National Sport in Nagoya. |
| December 9 | Sightseeing in Nagoya. |
| | Train journey to Kyoto. |

| December 10 | Delivers fifth public lecture at the Kyoto Civic Auditorium. |
| | Sightseeing in Kyoto, including the Imperial Palace. |
| December 11 | Train journey to Osaka. |
| | Delivers sixth public lecture at the Osaka Central Auditorium. |
| December 12 | Sightseeing in Osaka. |
| December 13 | Train journey to Kobe. |
| | Delivers seventh public lecture at the Kobe YMCA. |
| December 14 | Completes "Answer to Question on Religion" (*CPAE 2012*, Vol. 13, Doc. 398). |
| | Delivers impromptu lecture on "How I Created the Theory of Relativity" at a student reception in his honor at Kyoto University. |
| December 15 | Sightseeing in Kyoto. |
| December 16 | Excursion to Lake Biwa. |
| December 18 | Sightseeing in Nara. |
| December 19 | Train journey to Hiroshima. |
| December 20 | Arrives in Miyajima. |
| December 23 | Train journey to Moji. |
| December 24 | Travels to Fukuoka. |
| | Delivers eighth public lecture at the Hakata Daihaku Theater in Fukuoka. |

| December 25 | Banquet at the Kyushu Imperial University, Fukuoka. |
| | Return train journey to Moji. |
| | Plays violin at a children's Christmas party at the Moji YMCA. |
| December 26 | Sightseeing around Moji. |
| December 27 | Completes Preface for Japanese edition of his works (*CPAE 2012*, Vol. 13, Doc. 406). |
| | Brief cruise through the Straits of Shimonoseki. |
| December 29 | Departs Japan from Moji on board the S.S. *Haruna Maru*. |
| December 31 | Arrives in Shanghai. |

### 1923

| January | "Musings on My Impressions in Japan" (*CPAE 2012*, Vol. 13, Doc. 391) is published. |
| January 1 | Reception held by Shanghai Jewish community at home of S. Gatton in honor of Einstein. |
| | Discussion on relativity at the Shanghai Municipal Committee. |
| January 2 | Departs from Shanghai. |
| January 5 | Arrives in Hong Kong. |
| January 6 | Departs from Hong Kong. |
| January 10 | Arrives in Singapore. |

| | |
|---|---|
| January 12 | Departs from Singapore. |
| January 13 | Arrives in and departs from Malacca. |
| January 14 | Arrives in Penang. |
| January 15 | Arrives in Colombo. |
| | Excursion to Negombo. |
| January 22 | Completes "On the General Theory of Relativity" (*CPAE 2012*, Vol. 13, Doc. 425). |
| January 31 | Arrives in Suez, Egypt. |
| February | "Answer to Question on Religion" (*CPAE 2012*, Vol. 13, Doc. 398) is published. |
| February 1 | Arrives in Port Said. |
| | Train journey to Kantara. |
| | Crosses Suez Canal by ferry. |
| | Train journey toward Lod, Palestine. |
| February 2 | Arrives in Lod and welcomed by Zionist dignitaries. |
| | Changes trains and travels to Jerusalem. |
| | Arrives in Jerusalem and lodges with Sir Herbert Samuel. |
| February 3 | Tours Muslim and Jewish quarters of Old City of Jerusalem, including the Dome of the Rock, Al-Aqsa Mosque, and the Western Wall. |
| February 4 | Excursion to Jericho and Allenby Bridge. |

| February 5 | Tours western Jerusalem, including National Library. |
| February 6 | Tours Bezalel Art Academy. |
| | Official welcome by the Jewish community of Jerusalem at the Lämel School. |
| February 7 | Tours Christian quarter of Old City. |
| | Lectures at the future site of the Hebrew University on Mount Scopus in Jerusalem. |
| | Banquet in his honor at Government House. |
| February 8 | Travels by car to Tel Aviv. |
| | Reception at "Herzliya" Gymnasium in Einstein's honor. |
| | Official welcome at Tel Aviv City Hall; named honorary citizen of Tel Aviv. |
| | Tour of various infrastructure projects in Tel Aviv. |
| | Public reception in Einstein's honor at "Herzliya" Gymnasium. |
| February 9 | Attends session of General Federation of Labor's semiannual conference. |
| | Visits Mikve Israel agricultural school. |
| | Visits Rishon LeZion. |
| | Train journey to Haifa. |
| February 10 | Attends receptions in his honor at the Technion. |
| | Festive banquet in his honor at the Reali School. |

| | |
|---|---|
| February 11 | Tours the Reali School. |
| | Tours industrial projects in Haifa vicinity. |
| | Travels by car via Nazareth toward Sea of Galilee. |
| | Tours Nahalal moshav. |
| | Arrives in Migdal. |
| February 12 | Travels by car to Sea of Galilee. |
| | Tours Degania kibbutz. |
| | Arrives in Nazareth. |
| February 13 | Travels by car to Jerusalem. |
| | Lectures on relativity at Lämel School. |
| February 14 | Departs Jerusalem by train for Lod. Changes trains en route to Kantara. |
| | Arrives in Port Said. |
| February 16 | Departs on board the S.S. *Ormuz*. |
| February 21 or 22 | Disembarks in Toulon. |
| February 22 | Arrives in Barcelona by train from Marseille. |
| February 24 | Delivers his first lecture in Barcelona at the provincial government building. |
| February 25 | Tours Poblet Monastery outside of Barcelona and the town of L'Espluga de Francolí. |
| February 26 | Tours city of Terrassa. |
| | Delivers his second lecture in Barcelona at the provincial government building. |

| February 27 | Tours two innovative schools. |
| | Reception at Barcelona City Hall. Delivers speech to the municipality of Barcelona. |
| | Lectures at the Royal Academy of Sciences and Arts in Barcelona. |
| | Farewell banquet for Einstein. |
| February 28 | Tours the Industrial School of Barcelona. |
| March 1 | Departs Barcelona by train. |
| | Arrives in Madrid. |
| March 2 | Sightseeing drive through Madrid. |
| | Visits the Laboratory of Physical Research. |
| March 3 | First tour of Prado. |
| | Official welcome at Madrid City Hall. |
| | Delivers first public lecture at the Universidad Central. |
| | Banquet at the Palace Hotel. |
| March 4 | Delivers lecture to special session of the Royal Academy of Exact, Physical, and Natural Sciences presided over by King Alfonso XIII. |
| March 5 | Attends special session of the Mathematical Society. |
| | Delivers second public lecture at the Universidad Central. |
| March 6 | Excursion to Toledo. |
| | Nominated as a corresponding member of the Royal Academy of Sciences and Arts of Barcelona. |

| | |
|---|---|
| March 7 | Has audience with King Alfonso XIII and Queen Mother Maria Christina of Austria at the Royal Palace. |
| | Delivers third public lecture at the Universidad Central. |
| | Attends reception at German ambassador's residence in his honor. |
| March 8 | Receives honorary doctorate from the Universidad Central. |
| | Delivers fourth public lecture at the Madrid Athenaeum. |
| March 9 | Tours El Escorial and Mendoza Castle. |
| | Attends public tribute in his honor at the Residencia de Estudiantes at the Universidad Central. |
| March 10 | Second tour of Prado. |
| March 11 | Third tour of Prado. |
| March 12 | Travels by train to Zaragoza. |
| | Delivers first lecture at the Faculty of Medicine and Sciences at the University of Zaragoza. |
| March 13 | Tours Zaragoza. |
| | Official lunch held in his honor at the Centro Mercantile. |
| | Delivers second lecture at the Faculty of Medicine and Sciences at the University of Zaragoza. |
| | Attends banquet in his honor at German consul's residence. |

March 14       Departs Zaragoza for Barcelona.

March 15       Travels from Barcelona to Zurich.

March 21       Travels from Zurich to Berlin.

# Abbreviations

### Descriptive Codes

| | |
|---|---|
| AD | Autograph Document |
| AKS | Autograph Postcard Signed |
| ALS | Autograph Letter Signed |
| ALSX | Autograph Letter Signed Xerox |
| PLS | Printed Letter Signed |
| REPT | Reprint |
| TDS | Typed Document Signed |
| TLS | Typed Letter Signed |
| TTrL | Typed Transcript Letter |

### Location Symbols

| | |
|---|---|
| AEA | Albert Einstein Archives, Hebrew University of Jerusalem |
| CPT | Archives of the California Institute of Technology, Pasadena, California |
| DkKoNBA | Niels Bohr Archive, Copenhagen |
| EPPA | Einstein Papers Project Archives, California Institute of Technology, Pasadena, California |
| Es-BaACA | Royal Academy of Sciences and Arts of Barcelona Archives, Barcelona |
| GyBAr (B) | Deutsches Bundesarchiv, Berlin |
| GyBPAAA | Politisches Archiv des Auswärtigen Amtes, Berlin |
| GyBSA | Geheimes Staatsarchiv, Preußischer Kulturbesitz, Berlin (Dahlem) |

| | |
|---|---|
| IsJCZA | Central Zionist Archives, Jerusalem |
| IsReWW | Yad Chaim Weizmann (Weizmann Archives), Weizmann Institute, Rehovoth, Israel |
| JSeTU | Tohoku University Library, Sendai, Japan |
| JTDRO | Diplomatic Record Office, Tokyo |
| JTJA | The Japan Academy, Tokyo |
| JTNAJ | National Archives of Japan, Tokyo |
| NjP-L | Princeton University Library, Princeton, New Jersey |
| NNLBI | Leo Baeck Institute, New York |
| NNPM | The Pierpont Morgan Library, New York |
| SSVA | Kungliga Vetenskapsakademien, Stockholm |
| SzZuETH | Eidgenössische Technische Hochschule, Zurich |
| SzZuZB | Zentralbibliothek, Zurich |

# Notes

## Historical Introduction

1. For general overviews of the trip to the Far East, Palestine, and Spain, see *Grundmann 2004*, pp. 223–250; *Eisinger 2011*, pp. 21–71; and *Calaprice et al. 2015*, pp. 111–115.

2. See "Travel Diary Japan, Palestine, Spain, 6 October 1922–12 March 1923 [*CPAE 2012*, Vol. 13, Doc. 379, pp. 532–588].

3. See *Sugimoto 2001b*, pp. 12–133; *Rosenkranz 1999*; "Einstein's Travel Diary for Spain, 1923," in *Glick 1988*, pp. 325–326; and *Nathan and Norden 1975*, pp. 75–76.

4. On Einstein's trip to the United States in the spring of 1921, see *CPAE 2009*, Vol. 12, Introduction, pp. xxviii–xxxviii.

5. See "South American Travel Diary Argentina, Uruguay, Brazil," 5 March–11 May 1925 [*CPAE 2015*, Vol. 14, Doc. 455, pp. 688–708]; "Amerika-Reise 1930," 30 November 1930–15 June 1931 [AEA, 29 134]; "Travel diary for USA," 3 December 1931–4 February 1932 [AEA, 29 136]; "Reise nach Pasadena XII 1932," 10 December 1932–18 December 1932 [AEA, 29 138]; and "Travel Diary for Pasadena," 28 January 1933–16 February 1933 [AEA, 29 143].

6. See "Calculations on Back Pages of Travel Diary," ca. 9–22 January 1923 [*CPAE 2012*, Vol. 13, Doc. 418, pp. 670–694].

7. Indirect evidence for this can be derived from a letter Einstein wrote while on his trip to South America. On 15 April 1925, he wrote home from Buenos Aires: "What adventures I have had! You will read about it in my diary" (see Einstein to Elsa and Margot Einstein, 15 April 1925 [*CPAE 2015*, Vol. 14, Doc. 474]).

8. See *Sayen 1985*, p. 72.

9. See *Bailey 1989*, pp. 348–351.

10. See "Note for the files," 3 March 1980 [AEA, Helen Dukas Papers, Heineman Foundation file].

11. See James H. Heineman to Otto Nathan, 21 October 1980 [AEA, Helen Dukas Papers, Heineman Foundation file].

12. See Charles Hamilton Galleries Inc., "Certification," 8 July 1981, and John F. Fleming, untitled appraisal, 8 July 1981 [AEA, Helen Dukas Papers, Heineman Foundation file].

13. See Otto Nathan to James H. Heineman, 15 August 1981, and James H. Heineman to Otto Nathan, 20 August 1981 [AEA, Helen Dukas Papers, Heineman Foundation file].

14. On the various factors, see *Grundmann 2004*, pp. 180–183.

15. See see *CPAE 2009*, Vol. 12, Introduction, pp. xxviii–xxxviii.

16. He held a lecture cycle at the University of Zurich in January–February 1919; a series of lectures in Oslo and a lecture on relativity at the Technical University of Copenhagen in June 1920; he gave his inaugural lecture at the University of Leyden in October 1920; and he delivered lectures at the Urania in Prague and at the University in Vienna in January 1921 (see *CPAE 2004*, Vol. 9, Calendar, entry for 20 January 1919; *CPAE 2006*, Vol. 10, Calendar, entries for 15, 17, 18, and 25 June 1920 and 27 October 1920; and *CPAE 2009*, Vol. 12, Calendar, entries for 7, 8, 10, 11, and 13 January 1921).

17. In June 1921, he addressed Jewish students on the Hebrew University at the University of Manchester (see *CPAE 2009*, Vol. 12, Calendar, entry for 9 June 1921).

18. See *CPAE 2012*, Vol. 13, Chronology, entries for 3, 5–7 April 1922.

19. See *Kagawa 1920*.

20. See Sanehiko Yamamoto, "Fifteen Years of Kaizo," *Kaizo*, April 1934; and *Kaneko 2005*, p. 13.

21. See Jun Ishiwara, "Preface," in *Ishiwara 1923*.

22. See *Yokozeki 1956*.

23. See Jun Ishiwara to Einstein, 24 September 1921 [*CPAE 2009*, Vol. 12, Doc. 244].

24. See Koshin Morubuse to Einstein, before 27 September 1921 [*CPAE 2009*, Vol. 12, Doc. 245].

25. See Einstein to Elsa Einstein, 8 January 1921 [*CPAE 2009*, Vol. 12, Doc. 12].

26. See Einstein to Ilse Einstein, 9 November 1921 [*CPAE 2009*, Vol. 12, Doc. 292].

27. See Einstein to Jun Ishiwara, 6 December 1921 [*CPAE 2009*, Vol. 12, Doc. 312].

28. See Sanehiko Yamamoto to Einstein, 15 January 1922 [*CPAE 2012*, Vol. 13, Doc. 21].

29. See Einstein to Paul Ehrenfest, 15 March 1922 [*CPAE 2012*, Vol. 13, Doc. 87].

30. See Eintein to Jun Ishiwara, 27 March 1922 [*CPAE 2012*, Vol. 13, Doc. 118].

31. See Sanehiko Yamamoto to Einstein, between 12 July and 8 August 1922 [*CPAE 2012*, Vol. 13, Doc. 283].

32. See Uzumi Doi to Einstein, 27 May 1922 [*CPAE 2012*, Vol. 13, Doc. 206].

33. See Einstein to Koshin Morubuse, 27 September 1921 [*CPAE 2009*, Vol. 12, Doc. 246].

34. See Text 4 in the Additional Texts section of this volume.

35. See W. S. Ting, Chinese Embassy, Copenhagen to Einstein, 11 September 1920 [*CPAE 2006*, Vol. 10, Calendar, entry for 11 September 1920].

36. See Zhu Jia-hua to Einstein, 21 March 1922 [*CPAE 2012*, Vol. 13, Doc. 101].

37. See Einstein to Zhu Jia-hua, 25 March 1922 [*CPAE 2012*, Vol. 13, Doc. 111].

38. See Chenzu Wei to Einstein, 8 April 1922 [*CPAE 2012*, Vol. 13, Doc. 135].

39. See Einstein to Chenzu Wei, 3 May 1922 [*CPAE 2012*, Vol. 13, Doc. 177].

40. See Chenzu Wei to Einstein, 22 July 1922 [*CPAE 2012*, Vol. 13, Doc. 305].

41. On the trip to Palestine, see *Rosenkranz 2011*, pp. 139–180.

42. See Chaim Weizmann to Einstein, 7 October 1921 [*CPAE 2009*, Vol. 12, Doc. 259].

43. This invitation was a written one, yet it is not extant. For its existence, see Arthur Ruppin to Zionist Executive, 16 Oct. 1922 [IsJCZA, A126/542]. For Blumenfeld's notes of 12 October 1922, see Text 2 in the Additional Texts section of this volume.

44. Einstein was to travel to Batavia (Java) to show his gratitude to the joint Dutch-German expedition that had observed a solar eclipse there in one of the attempts to prove his theories (see Einstein to Paul Ehrenfest, 18 May 1922 [*CPAE 2009*, Vol. 12, Doc. 193]).

45. See Text 2 in the Additional Texts section of this volume.

46. See "Prof. Einstein besucht Palästina," *Zionistische Korrespondenz*, 6 October 1922; and "Einstein to visit Palestine," *Latest News and Wires through Jewish Correspondence Bureau News and Telegraphic Agency*, 10 October 1922.

47. See Weizmann to Einstein, 6 Oct. 1922 [*CPAE 2012*, Vol. 13, Doc. 380]; Ilse Einstein to Weizmann, 20 October 1922 [*CPAE 2012*, Vol. 13, Abs. 435]; and *Wasserstein 1977*, Introduction, note 15.

48. On the problematic issues that arose during Einstein's tour of the United States, see *CPAE 2009*, Vol. 12, Introduction, p. xxxiv.

49. Solomon Ginzberg, who had acted as Einstein's host during the U.S. tour.

50. Her name was actually Rosa Ginzberg; she was Solomon's wife.

51. See Arthur Ruppin to Zionist Executive, Jerusalem, 16 October 1922 [IsReWW].

52. See Arthur Ruppin to Chaim Weizmann, 16 October 1922 [IsReWW].

53. See Einstein to Fritz Haber, 6 October 1920 [*CPAE 2006*, Vol. 10, Doc. 162].

54. See Julio Rey Pastor to Einstein, 22 April 1920 [*CPAE 2004*, Vol. 9, Doc. 391]; *CPAE 2004*, Vol. 9, Calendar, entry for 28 April 1920, and *CPAE 2009*, Vol. 12, Calendar, entry for 1 July 1921.

55. See Einstein to Heinrich Zangger, 18 June 1922 [*CPAE 2012*, Vol. 13, Doc. 241].

56. For details on Rathenau's assassination, see *Sabrow 1994a* and *Sabrow 1999*.

57. See *Sabrow 1994b*, pp. 157–169.

58. See Einstein to Mathilde Rathenau, after 24 June 1922 [*CPAE 2012*, Vol. 13, Doc. 245].

59. See "In Memorium Walther Rathenau," August 1922 [*CPAE 2012*, Vol. 13, Doc. 317].

60. See Mileva Einstein-Marić to Einstein, after 24 June 1922 [*CPAE 2012*, Vol. 13, Doc. 248].

61. Friedrich Sternthal to Einstein, 28 June 1922 [*CPAE 2012*, Vol. 13, Doc. 253].

62. Hermann Anschütz-Kaempfe to Einstein, 25 June 1922 [*CPAE 2012*, Vol. 13, Doc. 250].

63. Einstein to Hermann Anschütz-Kaempfe, 1 July 1922 [*CPAE 2012*, Vol. 13, Doc. 257].

64. Einstein had previously expressed his desire to live in the countryside and leave Berlin, which he found "nerve-racking" (see Einstein to Elsa Einstein, 14 September 1920 [*CPAE 2006*, Vol. 10, Doc. 149]). However, that same month, Einstein had reassured his closest political confidant, Konrad Haenisch, that "Berlin is the place in which I am most deeply rooted through personal and professional ties." Therefore, he would not leave Berlin unless "external circumstances forced him to do so" (see Einstein to Konrad Haenisch, 8 September 1920 [*CPAE 2006*, Vol. 10, Doc. 137]).

65. Einstein to Marie Curie-Skłodowska, 11 July 1922 [*CPAE 2012*, Vol. 13, Doc. 275].

66. Einstein to Max von Laue, 12 July 1922 [*CPAE 2012*, Vol. 13, Doc. 278].

67. Einstein to Hermann Anschütz-Kaempfe, 12 July 1922 [*CPAE 2012*, Vol. 13, Doc. 276].

68. Elsa's postscript to Einstein to Hermann Anschütz-Kaempfe, 16 July 1922 [*CPAE 2012*, Vol. 13, Doc. 292].

69. Einstein to Max Planck, 6 July 1922 [*CPAE 2012*, Vol. 13, Doc. 266].

70. See Hugo Bergmann to Einstein, 22 October 1919 [*CPAE 2004*, Vol. 9, Doc. 147].

71. See *Bergman 1919*, pp. 4–5. This article was published in Hebrew. However, it is quite likely that it was also published in German at the time.

72. See Einstein to Paul Epstein, 5 October 1919 [*CPAE 2004*, Vol. 9, Doc. 122].

73. See text of diary, this volume, entry for 13 October 1922.

74. Ibid, entry for 14 October 1922.

75. On the image of the "allegedly uncorrupted, authentic Orient" in Karl May's writings, see *Krobb 2014*, p. 14.

76. See text of diary, this volume, entry for 1 February 1923.

77. Ibid, entry for 28 October 1922. Einstein did not distinguish between the Indian and the Sinhalese inhabitants of Colombo.

78. Ibid, entry for 19 January 1923.

79. Einstein to Paul Ehrenfest, 22 March 1919 [*CPAE 2004*, Vol. 9, Doc. 10].

80. Einstein to Emil Zürcher, 15 April 1919 [*CPAE 2004*, Vol. 9, Doc. 23].

81. Einstein to Heinrich Zangger, 24 December 1919 [*CPAE 2004*, Vol. 9, Doc. 233].

82. See text of diary, this volume, entry for 2 November 1922.

83. Ibid, entry for 3 November 1922.

84. Ibid, entry for 10 November 1922.

85. Ibid.

86. Ibid, entry for 14 November 1922.

87. Ibid, entry for 1 January 1923.

88. See *Clifford 2001*, p. 133–134.

89. Ibid.

90. Ibid, p. 132.

91. See Einstein to Maximilian Pfister, 28 August 1922 [*CPAE 2012*, Vol. 13, Doc. 331].

92. On the planned lecture tour in China and Einstein's visits to Shanghai, see *Hu 2005*, pp. 66–79.

93. See Yuanpei Cai to Einstein, 8 December 1922 [*CPAE 2012*, Vol. 13, Doc. 392].

94. See Einstein to Yuanpei Cai, 22 December 1922 [*CPAE 2012*, Vol. 13, Doc. 403].

95. See Einstein to Maurice Solovine, 18 March 1909 [*CPAE 1993*, Vol. 5, Doc. 142].

96. See Einstein to Jakob Laub, 4 November 1910 [*CPAE 1993*, Vol. 5, Doc. 231].

97. See Einstein to Ayao Kuwaki, 28 December 1920 [*CPAE 2006*, Vol. 10, Doc. 246].

98. See Einstein to Ilse Einstein, 9 November 1921 [*CPAE 2009*, Vol. 12, Doc. 292].

99. See Einstein to Elsa Einstein, 20 November 1921 [*CPAE 2009*, Vol. 12, Doc. 303].

100. See *Lambourne 2005*, p. 174.

101. See *Hashimoto 2005*, p. 104.

102. See *Kaneko 1984*, pp. 51–52.

103. See Text 4 in the Additional Texts section of this volume.

104. See *Jansen 1989*, pp. 147–148.

105. See *Kaneko 1987*, p. 354, and *Bellah 1972*, p. 109.

106. See *Gordon 2003*, pp. 161–180.

107. On the visit to Japan, see *Ezawa 2005*; *Jansen 1989*; *Kaneko 1981, 1984, 1987*, and *2005*; *Nisio 1979*; *Okamoto 1981*; and *Sugimoto 2001a, 2001b*.
108. See text of diary, this volume, entry for 8 October 1922.
109. Ibid, entry for 10 October 1922.
110. Ibid, entry for 31 October 1922.
111. Ibid, entries for 31 October and 3 November 1922.
112. Ibid, entries for 17–18 November 1922.
113. Ibid, entry for 5 December 1922.
114. Ibid.
115. Ibid, entry for 7 December 1922.
116. Ibid, entry for 10 December 1922.
117. See Text 6 in the Additional Texts section of this volume.
118. See Text 4 in the Additional Texts section of this volume.
119. See *Kalland and Asquith 1997*, pp. 1–2, and *Craig 2014*, p. 3.
120. Ibid, p. 5.
121. Ibid, p. 6.
122. See text of diary, this volume, entry for 25 November 1922.
123. See Text 4 in the Additional Texts section of this volume.
124. See Text 11 in the Additional Texts section of this volume.
125. See *Hashimoto 2005*, p. 121.
126. See text of diary, this volume, entry for 24 November 1922.
127. Ibid, entry for 25 November 1922.
128. See Text 4 in the Additional Texts section of this volume.
129. Ibid.
130. See *Hashimoto 2005*, p. 118.
131. See *Johnson 1993*, p. 138.
132. See *Neumann and Neumann 2003*, p. 187.
133. See Wilhelm Solf to German Foreign Ministry, 3 January 1923 (GyBPAAA/R 64882,).
134. See Text 7 in the Additional Texts section of this volume.
135. On the *Yishuv* during this period, see *Eliav 1976, Lissak 1993, Malamat et al. 1969*, pp. 272–288, and *Porat and Shavit 1982*.
136. See *Yapp 2003*, pp. 214, 217.
137. See *Kaiser 1992*, pp. 261–262, 265.
138. See *Metzler and Wildt 2012*, p. 189, and *Saposnik 2006*, pp. 1106, 1111–1112.
139. See text of diary, this volume, entry for 2 February 1923.
140. Ibid, entry for 3 February 1923.
141. See, e.g., *Friedman 1977*.
142. See *Kaiser 1992*, p. 271.
143. See text of diary, this volume, entry for 4 February 1923.
144. See Text 17 in the Additional Texts section of this volume.
145. On this issue, see, e.g., *Goldstein 1980*.

146. See "Notes in Palestine," [*CPAE 2012*, Vol. 13, Appendix G].

147. See text of diary, this volume, entry for 12 February 1923.

148. See Text 17 in the Additional Texts section of this volume.

149. See Einstein to Maurice Solovine, [20 May] 1923 [*CPAE 2015*, Vol. 14, Doc. 34].

150. See *Rosenkranz 2011*, p. 84.

151. See Text 17 in the Additional Texts section of this volume, and Einstein to Maurice Solovine, [20 May] 1923 [*CPAE 2015*, Vol. 14, Doc. 34].

152. See *Ben-Arieh 1989*.

153. Ibid.

154. See text of diary, this volume, entry for 12 February 1923.

155. Ibid, entry for 3 February 1923.

156. See Einstein to Ilse Einstein, 7 October 1920, and Ilse Einstein to Einstein, 10 October 1920 [*CPAE 2006*, Vol. 10, Docs. 165 and 173].

157. On Einstein's trip to Spain, see *Glick 1988*, *Roca Rossell 2005*, *Sánchez Ron and Romero de Pablos 2005*, and *Turrión Berges 2005*.

158. Even though the Spanish press reported a great deal of what Einstein allegedly uttered during his three weeks in Spain, it would undermine the methodology of this edition to consider those reports as authentic statements made by Einstein.

159. See text of diary, this volume, entry for 5 March 1923.

160. Ibid, entry for 7 March 1923.

161. Ibid, entry for 8 March 1923.

162. Ibid, entry for 22–28 February 1923.

163. Ibid, entries for 6 and 9 March 1923.

164. See Einstein to Paul Ehrenfest, 22 March 1919 [*CPAE 2004*, Vol. 9, Doc. 10].

165. See text of diary, this volume, entry for 14 November 1923.

166. Ibid, entry for 1 January 1923.

167. Ibid, entry for 28 October 1922.

168. See *Poiger 2005*, p. 121.

169. See *Fuhrmann 2011*, p. 126.

170. See text of diary, this volume, entry for 10 November 1922.

171. See *Root 2013*, p. 184.

172. See text of diary, this volume, entry for 19 January 1923.

173. Ibid, entry for 2 February 1923.

174. For such a study, see *Wiemann 1995*, p. 99; for the quote, see *Germana 2010*, p. 81.

175. See *Wilke 2011*, p. 291.

176. See *Pratt 1992*, p. 4.

177. See *Sachs 2003*, p. 117.

178. See *Lubrich 2004*, pp. 34, 37.

179. See *Said 1978*, p. 3.

180. For some examples of discussions of Said's theories, see *Dirlik 1996*, *Foster 1982*, *Lary 2006*, and *Marchand 2001*.

181. See *Foster 1982*, p. 21; *Jackson 1992*, p. 247; *Lary 2006*, p. 3; *Mudimbe-Boyi 1992*, p. 31; and *Wiemann 1995*, pp. 99–102.

182. See *Wiemann 1995*, p. 100.

183. See *Saposnik 2006*, pp. 1107–1108.

184. See *Aschheim 1982*, p. 187.

185. See *Saposnik 2006*, p. 1109.

186. Ibid, p. 1111.

187. See *Winteler-Einstein 1924*, pp. 25–26.

188. See Einstein to Elsa Einstein, 7 August 1917 [*CPAE 2006*, Vol. 8, Doc. 369b, in Vol. 10], and Einstein to Heinrich Zangger, 8 August 1917 [*CPAE 2006*, Vol. 8, Doc. 370a, in Vol. 10].

189. See Einstein to Luise Karr-Krüsi, 6 May 1919 [*CPAE 2004*, Vol. 9, Doc. 35a, in Vol. 13].

190. See text of diary, this volume, entries for 10 October and 2 November 1922, and 3 and 15 February 1923.

191. Ibid, entry for 6 October 1922.

192. See *Kaplan 1997*, p. 22.

193. See text of diary, this volume, entry for 14 November 1922.

194. Ibid, entry for 19 January 1923.

195. See *Kaplan 1997*, p. 6.

196. The phrase was coined by film critic Laura Mulvey in 1975 (see *Kaplan 1997*, p. 22).

197. See text of diary, this volume, entry for 31 December 1922.

198. Ibid, entry for 14 January 1923.

199. Ibid, entry for 19 January 1923.

200. Ibid, entry for 4 February 1923.

201. Ibid, entry for 14 February 1923.

202. See *Pratt 1992*, p. 7.

203. The phrase is borrowed from the African-American sociologist and civil rights leader W. E. B. Du Bois (see *Kaplan 1997*, pp. 8–10).

204. See *Mudimbe-Boyi 1992*, p. 28.

205. See *Seth and Knox 2006*, pp. 4–5, 214.

206. Ibid, p. 6.

207. See *Mudimbe-Boyi 1992*, p. 27.

208. See *Pratt 1985*, p. 139.

209. See *Kretschmer 1921*.

210. See *Leerssen 2000*, pp. 280–284.

211. See *Rosenkranz 2011*, pp. 261–262.

212. See *Doron 1980*, pp. 390–391.

213. See *Weiss 2006*, p. 51.

214. See *Doron 1980*, p. 391, and *Lipphardt 2016*, p. 112.

215. See *Weiss 2006*, pp. 51–52.

216. See *Weiss 2006*, p. 58, and *Gelber 2000*, p. 126.

217. See *Doron 1980*, p. 392, and *Niewyk 2001*, pp. 105–107

218. See *Doron 1980*, p. 398, note 26.

219. See *Falk 2006*, pp. 140–141.

220. See *Doron 1980*, p. 404, and *Hambrock 2003*, p. 52.

221. See *Doron 1980*, p. 412.

222. See ibid., p. 412, and *Niewyk 2001*, p. 130.

223. See *Niewyk 2001*, p. 130. See also *Doron 1980*, p. 422, and *Gelber 2000*, pp. 125–126.

224. See *Niewyk 2001*, p. 131.

225. See *Miles and Brown 2003*, p. 10.

226. Ibid, p. 85.

227. See *Miles 1982*, p. 157, quoted in *Miles and Brown 2003*, p. 100.

228. See *Miles and Brown 2003*, p. 103.

229. Ibid, p. 104.

230. See "On the Questionnaire Concerning the Right of National Self-Determination," July 1917–before 10 March 1918 [*CPAE 2002*, Vol. 6, Doc. 45a, in Vol. 7].

231. See "Assimilation and Anti-Semitism," 3 April 1920 [*CPAE 2002*, Vol. 7, Doc. 34].

232. See Einstein to Central Association of German Citizens of the Jewish Faith, 5 April 1920 [*CPAE 2004*, Vol. 9, Doc. 368].

233. See Einstein to Emil Starkenstein, 14 July 1921 [*CPAE 2009*, Vol. 12, Doc. 181].

234. On his encounter with the American Jewish community in spring 1921, see *CPAE 2009*, Vol. 12, Introduction, pp. xxxi–xxxiv.

235. See text of diary, this volume, entry for 10 November 1922.

236. Ibid, entry for 3 February 1923.

237. See Einstein to [Paul Nathan], 3 April 1920 [*CPAE 2004*, Vol. 9, Doc. 306].

238. See Illustration 18.

239. The question of whether other German Jewish and Zionist intellectuals expressed views of the alleged inferiority of other peoples in their *private* writings does not seem to have been studied by historians. It goes beyond the scope of this introduction to carry out such a study.

240. See *Miles and Brown 2003*, p. 104.

241. See *Youngs 2013*, p. 102.

242. See *Confino 2003*, p. 326.

243. See *Koshar 1998*, p. 325–326.

244. See *Clifford 2001*, p. 129.

245. See *Selwyn 1996*, p. 21.

246. See *Nünning 2008*, p. 16.

247. See *Walton 2009*, p. 117.

248. See *Keitz 1993*, p. 187.

249. See text of diary, this volume, entries for 9–10 October 1922.

250. See *Mansfield 2006–2007*, pp. 706–707, 711.

251. See *Kisch 1938*, pp. 29–31.

252. Personal communication with Barbara Wolff, AEA, 25 February 2008.

253. See, e.g., *Rosenkranz 2011*, p. 84.

254. See *Jokinen and Veijola 1997*.

255. See text of diary, this volume, entry for 6 October 1922.

256. Ibid, entry for 2 November 1922.

257. Ibid, entry for 7 December 1922.

258. Ibid, entry for 11 December 1922.

259. Einstein was inconsistent in how he spelled Elsa's name in the diary.

260. Ibid, entry for 15 February 1923.

261. Ibid, entry for 18 November 1922.

262. Ibid, entry for 29 November 1922.

263. Ibid, entry for 25 December 1922.

264. Ibid, entry for 7 February 1923.

265. Ibid, entry for 7 March 1923.

266. See Text 13 in the Additional Texts section of this volume.

267. See Text 6 in the Additional Texts section of this volume.

268. See *Einstein 1923b*.

269. See Text 14 in the Additional Texts section of this volume.

270. See Christopher Aurivillius to Einstein, 10 November 1922 [*CPAE 2012*, Vol. 13, Doc. 384]. On the proposals to award the prize to Einstein in earlier years, see, e.g., *Friedman 2001*, pp. 133–138.

271. See Svante Arrhenius to Einstein, on or before 17 September 1922 [*CPAE 2012*, Vol. 13, Doc. 359].

272. See Max von Laue to Einstein, 18 September 1922 [*CPAE 2012*, Vol. 13, Doc. 363].

273. See Einstein to Svante Arrhenius, 20 September 1922 [*CPAE 2012*, Vol. 13, Doc. 365].

274. See *Grundmann* 2004, pp. 180–182.

275. See *Renn 2013*, p. 2577.

276. See Einstein to Fritz Haber, 6 October 1920 [*CPAE 2006*, Vol. 10, Doc. 162].

277. See *CPAE 2009*, Vol. 12, Introduction, pp. xxxii, and, Calendar, entry for 9 June 1921.

278. For a collection of studies on the reception of relativity in various countries, see *Glick 1987*.

279. See *Hu 2007*, pp. 541–542.

280. See *Kaneko 1987*, p. 363.

281. Ibid, pp. 353–354.

282. See *Glick 1987*, p. 392.

283. Ibid, p. 362.

284. Ibid, p. 363.

285. Ibid, p. 354.

286. Ibid, pp. 372–374.

287. See text of diary, this volume, note 151, and "Preface for the Japanese edition of Georg Nicolai's *Biologie des Krieges*," 10 December 1922 [*CPAE 2012*, Vol. 13, Doc. 394, note 3].

288. See Text 15 in the Additional Texts section of this volume, note 66.

289. See, e.g., his comment in reaction to a political appeal, in which he argued against "the pointless exacerbation of antagonisms, which are, in themselves, necessary and productive" (see Einstein to Workers International Relief, 28 February 1926, *CPAE 2018*, Vol. 15, Doc. 206]).

290. See Text 15 in the Additional Texts section of this volume.

291. See *Glick 1987*, p. 353.

292. See Wilhelm Solf to German Foreign Ministry, 3 January 1923 [GyBSA, I. HA, Rep. 76 Vc, Sekt. 1, Tit. 11, Teil 5c, Nr. 55, Bl. 157–158], and *Neumann and Neumann 2003*, p. 187.

293. See "Prof. Einstein's lecture on Mt. Scopus" and untitled article by Aharon Czerniawski, *Ha'aretz*, 11 February 1923; and "The Einstein Theory," *The Palestine Weekly*, 9 February 1923, pp. 83–84.

294. See *Berkowitz 2012*, p. 223.

295. See "The reception for Prof. Einstein at the Lemel School," *Do'ar Hayom*, 8 February 1923; "The Opening of the Hebrew College," *Do'ar Hayom*, 9 February 1923; and "Prof. Einstein's lecture on Mt. Scopus," *Ha'aretz*, 11 February 1923.

296. See Chaim Weizmann to Einstein, 4 February 1923 [*CPAE 2012*, Vol. 13, Doc. 427].

297. See "Prof. Einstein's lecture on Mt. Scopus," *Ha'aretz*, 11 February 1923; "Einstein in Eretz Yisrael," *Aspeklarya* No. 12, 1 February 1923, p. 5; and untitled article, *Do'ar Hayom*, 9 February 1923.

298. See Illustration 34.

299. See *Glick 1987*, pp. 231–234, 243–244.

300. Ibid, pp. 252–258, 395; and *Renn 2013*, pp. 2583–2585.

301. See *Glick 1987*, p. 393.

302. See *Renn 2013*, p. 2581.

303. See *Glick 1988*, p. 70.

304. See *Renn 2013*, p. 2583.

305. See *Patiniotis and Gavroglu 2012*, p. 1.

306. See *Pratt 1992*, p. 6.

307. See text of diary, this volume, entry for 10 October 1922.

308. Some instances of Einstein's biological worldview and of his views on genetics are as follows. In 1917, Einstein expressed remorse and self-reproach for having fathered children with his first wife Mileva, "a physically and morally inferior person." Yet at the same time, he admitted that his own family lacked a high-quality genetic pedigree. In March 1917, he raised the possibility of imitating "the methods of the Spartans" to deal with his son Eduard's alleged genetic inferiority. A year later, he differentiated between "valuable people" (i.e., those with superior intellect) and those who were less valuable (i.e., "unimaginative average [people]") and who were therefore more expendable in war. A further instance of Einstein defining which human lives he found "valuable" and which not can be seen in the manner in which he related to his close friend Paul Ehrenfest's youngest son Wassily being diagnosed with Down syndrome. He approved of the plan "to hand the child over to impersonal care," and added that "valuable people should not be sacrificed to hopeless causes" (see Einstein to Heinrich Zangger, 16 February 1917 [*CPAE 2006*, Vol. 8, 299a, in Vol. 10]; Einstein to Otto Heinrich Warburg, 23 March 1918 [*CPAE 1998*, Vol. 8, Doc. 491]; and Einstein to Paul Ehrenfest, on or after 22 August 1922 [*CPAE 2012*, Vol. 13, Doc. 329]).

309. See *Yamamoto 1934*.

310. See Einstein to Mileva Marić, 14? August 1900 [*CPAE 1987*, Vol. 1, Doc. 72].

311. See *Isaacson 2008*, p. 289.

312. On the limits of Einstein's humanism, see *Rosenkranz 2011*, pp. 266–267.

## Travel Diary: Japan, Palestine, Spain,
## 6 October 1922–12 March 1923

1. AD (NNPM, MA 3951). [AEA, 29 129]. Published as "Travel Diary Japan, Palestine, Spain" in *CPAE 2012*, Vol. 13, Doc. 379, pp. 532–588. Excerpts published in *Nathan and Norden 1975*, pp. 75–76; "Einstein's Travel Diary for Spain, 1923," in *Glick 1988*, pp. 325–326; *Rosenkranz 1999*; and Sugimoto *2001b*, pp. 12–133. The document presented here comprises part of a notebook that measures 22.7 × 17.5 cm and consists of 182 lined pages. The notebook includes 81 lined pages of travel diary entries, followed by 82 blank lined pages and 19 lined pages and one unlined page of calculations (see *CPAE 2012*, Vol. 13, Doc. 418). These calculations were written on the verso of this document, i.e., at the back end of the travel diary and upside down in relation to the diary entries. The pages of the notebook have been numbered by the NNPM. On the inside of the front flyleaf, Einstein's secretary Helen Dukas has noted "Reise nach Japan Palestine Spanien 6. Oktober 1922–12. 3. 23." ("Trip to Japan Palestine Spain 6 October 1922–12 March 1923.") The diary entries appear on pp. 1 to 41v. Page 39v is blank. The diary entries are written in ink, with the exception of pp. 1, 5v, and 6, which are written entirely in pencil, and pp. 1v, 2v, 3, 3v, 4v, and 5, which are written partially in pencil. In this translation, deletions in the original text that are deemed significant are placed within angle brackets.

2. Michele Besso (1873–1955) was a Swiss-Italian engineer and close friend of Einstein. He was employed at the Swiss Patent Office in Bern and lectured on patent law at the Polytechnic in Zurich.

3. Lucien Chavan (1868–1942) was a friend of Einstein and a retired Swiss-French electrotechnician in Bern.

   Einstein had departed Berlin on 3 October and had visited his sons in Zurich on 3–4 October (see Edgar Meyer to Paul Epstein, 4 October 1922 [CPT, Paul Epstein Collection, folder 5.60]).

4. Elsa Einstein (1876–1936) was Einstein's second wife.

5. The S.S. *Kitano Maru* was owned by the Nippon Yusen Kaisha shipping company. It was built in 1909 and sailed the route from Antwerp to Yokohama. In 1942, the ship was sunk by a Japanese mine in the Lingayen Gulf in the Philippines.

6. This was probably Hayari Miyake (1866–1945), professor of surgery at Kyushu Imperial University. As Miyake was short in stature, Einstein may have initially misjudged his age. He had been in Europe to inspect medical institutions and surgical equipment on behalf of the Japanese government. He had also collected signatures among European surgeons for a petition to the International Surgical Association against its boycott of former Axis countries.

The physician from Munich may have been Ernst Ferdinand Sauerbruch (1875–1951), professor of medicine at the University of Munich, whom Miyake visited while in Europe. However, in contrast to Einstein's reference to Miyake's expulsion, Sauerbruch was allegedly one of the many German surgeons who came to see Miyake off in Marseille and even introduced him to Einstein (see *Hiki 2009*, p. 13).

7. Hayari Miyake. He had studied medicine at the University of Breslau.

8. *Kretschmer 1921*.

9. *Bergson 1922*.

10. In Riemannian geometry, the direction of a vector need not be preserved when the vector is transported around a closed curve. In Hermann Weyl's unified field theory of gravity and electromagnetism, the electromagnetic gauge field gets included in the geometry when local conformal invariance of the line element *ds* is demanded; as a consequence, the magnitude of a vector also no longer needs to be preserved when it is parallel transported around a closed curve (see *Weyl 1918*).

11. A small island off the north coast of Sicily in the Tyrrhenian Sea. It contains one of the three active volcanoes in Italy.

12. Paul Oppenheim (1885–1977) was a German-Jewish chemist with N.M. Oppenheim Nachfolger.

13. Viscount Kikujiro Ishii (1866–1945), Japanese ambassador to France and a Japanese representative at the League of Nations.

14. Through the Suez Canal.

15. Presumably the Great Bitter Lake, the saltwater lake between the northern and southern parts of the Suez Canal.

16. A mountain in the Swiss Plateau mountain range, on the outskirts of Zurich.

17. In a memorandum, Hayari Miyake recalled that Einstein believed he might have colon cancer. However, Miyake reassured Einstein that this was not the case (see *Kaneko 1981*, vol. 1, p. 178).

18. Cape Guardafui on the Horn of Africa.

19. Hayari Miyake.

20. Possibly the Kelaniya Raja Maha Vihara Temple, the foremost Buddhist temple in the vicinity of Colombo.

21. Probably the Pettah district in Colombo.

22. Emperor Yoshihito (1879–1926), who reigned during the Taishō era from 1912 to 1926. The menu of the banquet in honor of the emperor is available (see Illustration 8 [AEA, 36 454]).

23. "Banzai" is a traditional Japanese exclamation meaning "long life." The Japanese national anthem was "Kimigayo" ("The Emperor's Reign").

24. This was a performance of *joruri* or *gidayu*, a traditional Japanese dramatic narrative chanted to the accompaniment of the *samisen*, a string instrument.

25. Shin (Noboru) Sakuma (1893–1987), third secretary at the Japanese Embassy in Berlin. He had arranged the meeting between Einstein and Koshin Murobuse, the Europe correspondent of the *Kaizo* journal, in Berlin in September 1921 (see *Kaneko 1981*, vol. 1, p. 63).

26. Alfred Montor (1878–1950), a diamond merchant, Anna Montor (1886–1945), Max Montor (1872–1934). On Einstein's arrival, see *The Straits Times*, 3 November 1922.

27. Chaim Weizmann (1874–1952) was the president of the Zionist Organisation in London. He telegraphed the Singapore Zionist Society in mid-October to request a reception for the Einsteins in Singapore, at which funds for the Hebrew University were to be raised. Weizmann requested that Einstein be informed by cable on his arrival in Colombo of the planned reception (see Chaim Weizmann to Singapore Zionist Society, 12 October 1922 [IsReWW], and Chaim Weizmann to Manasseh Meyer, 12 October 1922 [IsJCZA, Z4/2685]).

28. The address of the Jewish community read by Montor was written by D. Kitovitz (see [C. R. Ginsburg], Singapore Zionist Society to Israel Cohen, Zionist Organisation, London, 9 November 1922 [IsJCZA, Z4/2685]). Montor welcomed Einstein on behalf of the Jewish community, praised his contributions to science, and expressed the wish that he take up the directorship of the Hebrew University. Einstein's speech was greeted with cheers. For the entire text of Montor's speech, see *The Straits Times*, 3 November 1922. For the text of Einstein's speech, see Text 3 in the Additional Texts section of this volume. *Meyer's Konversations-Lexikon* was a major German encyclopedia.

29. Manasseh Meyer (1846–1930) was the leader of the Singapore Jewish community and a prominent philanthropist. Croesus, ancient king of Lydia, was renowned for his legendary wealth.

30. The Chesed-El Synagogue was built on the grounds of Meyer's residence in 1905.

31. Hendrik A. Lorentz (1853–1928) was professor of theoretical physics at the University of Leyden.

32. Mozelle Nissim (1883–1975). This is a reference to the saying "beauty is the oldest nobility in nature," which originated in *Kotzebue 1792*, p. 59.

33. See Illustration 9.

34. The reception took place at 5 p.m. at Belle Vue, Oxley Rise in Singapore. At the event, "[a]ll communities and creeds were represented." Some 300 guests attended the gathering, including leading members of the Jewish community and the Anglican bishop (see *The Straits Times*, 31 October and 3 November 1922, and *Israel's Messenger*, 1 December 1922).

35. Charles James Ferguson-Davie (1872–1963), the Anglican bishop of Singapore.

36. The press reported that the banquet was attended by forty guests and hosted by Mozelle Nissim at the Meyer mansion (see *Israel's Messenger*, 1 December 1922). The guests were entertained by Braddon's orchestra (see *The Straits Times*, 3 November 1922).

37. The English translation hardly does justice to the German original of this idiom and its racist phrasing: "negerte lustig drauf los"—literally, "merrily niggered away."

38. A week following Einstein's visit, C. R. Ginsburg, the honorary secretary of the Singapore Zionist Society, informed the Zionist Organisation in London that Meyer had donated 500 pounds sterling to the Hebrew University, "partly as a result of a private interview with Professor Einstein who explained the object

of the collection, but chiefly I think as a result of a charming letter just received
by Mr. Meyer from Dr. Weizmann." The rest of the Jewish community donated
250 pounds sterling (see [C. R. Ginsburg], Singapore Zionist Society to Israel
Cohen, Zionist Organisation, London, 9 November 1922 [IsJCZA, Z4/2685]).
It was hoped that additional funds could be raised on the occasion of Einstein's
return visit following his tour of Japan (see [Israel Cohen], general secretary of
the Zionist Organisation, London, to David Kitovitz, 12 December 1922 [IsJCZA,
Z4/2685]).

39. The Peak, at 552 meters, is the highest elevation on Hong Kong Island.

40. According to press reports, when news first broke of Einstein's pending arrival
in Hong Kong, arrangements were made for a lecture at the Jewish Recreation
Club. However, after his arrival, Einstein asked that no receptions or lectures be
held. The press speculated that one possible reason he did not want to appear in
public during his brief stay was the proximity of his visit to Armistice Day. The
only planned event was a tour of Repulse Bay (see *South China Morning Post*, 10
November 1922).

41. One of the businessmen was probably named Gobin (see note 195).

42. Possibly the Repulse Bay Hotel.

43. Most likely a reference to the successful Chinese seamen's strike in early 1922 (see
*Butenhoff 1999*, p. 50).

44. The informal reception was held at the Jewish Recreation Club (see *Israel's Messenger*, 1 December 1922).

45. The University of Hong Kong, founded in 1911.

46. The S.S. *Kitano Maru* docked at Shanghai's Wayside Wharf.

47. Morikatsu Inagaki (1893–?) was a staff member of Kaizo-Sha and chief secretary
of the newly established Japanese Association of the League of Nations. He had
been asked by Sanehiko Yamamoto, president of Kaizo-Sha, to serve as Einstein's
guide and interpreter during his tour of Japan. His role was to interpret all of Einstein's speeches and daily conversations, but not Einstein's scientific lectures (see
*Ishiwara 1923*, "Preface," p. 10; *Kaneko 1981*, vol. 1, p. 14; and *Kaneko 1984*, p. 70). His
wife was Tony Inagaki, who was German-born.

48. Fritz Thiel (1863–1931). Three reports by Thiel on Einstein's visits to Shanghai
are extant. Following the first visit, he informed the German Foreign Ministry that
he had delivered several invitations to Einstein from Japan and the Malay Archipelago and invited him to breakfast at his private residence. However, as Einstein
was "preoccupied to such an extent by a Japanese sent to Shanghai to greet him, I
had to retreat." Nevertheless, Einstein informed Thiel that he could not accept any
scientific engagements unless they were coordinated with Kaizo-Sha, with whom
he was under contractual obligation. Thiel urged Einstein that "he must not pass
over" the German communities and the associations dedicated to the fostering of
German-Japanese scientific and cultural ties "without taking any notice of them."
Thiel rejected Einstein's claim that the journal held a monopoly on Einstein's

"entire personality." He warned Einstein that he would not have time for "the fulfillment of national duties" or for personal recreation. In response, Einstein reassured Thiel that he would take these viewpoints into consideration. He informed Thiel that he felt obligated to accept an invitation to visit Batavia (present-day Jakarta) and was therefore doubtful whether he would be able to follow through with the invitation to hold a lecture series in China.

Following the second visit, in an effort to repudiate rumors that the German community in Shanghai had snubbed Einstein for anti-Semitic reasons, Thiel informed the German Foreign Ministry that he had attempted to arrange for a lecture by Einstein at the Tongji University School of Engineering but had not received any reply. A few days prior to Einstein's return to Shanghai, the local German association received a postcard from Elsa Einstein declining an invitation to a reception. Moreover, when he learned that Einstein would allegedly be giving a talk on relativity to a closed reception of the Jewish community, Thiel decided to ignore Einstein's second visit to Shanghai (see Fritz Thiel to German Foreign Ministry, 13 November 1922 [GyBPAAA/R 9208/3508 Deutsche Botschaft China]; Fritz Thiel to Hubert Knipping, 28 November 1922; and Fritz Thiel to German Foreign Ministry, 6 January 192[3] [GyBPAAA/R 64677]).

49. Maximilian Pfister (1874–?), professor of internal medicine at the Tung Chi School of Medicine in Shanghai; Anna Pfister-Königsberger (1876–?).

50. It was on his arrival in Shanghai that Einstein learned that he had been awarded the Nobel Prize for Physics. He first received a telegram and a letter from the secretary of the Royal Swedish Academy of Sciences and was subsequently informed of the award by the Swedish consul-general (see Christopher Aurivillius to Einstein, 10 November 1922 [*CPAE 2012*, Vol. 13, Docs. 384 and 385]). The press reported that he "expressed great pleasure over being awarded the Nobel Prize" (see *The China Press*, 14 November 1922, and *Min Guo Ri Bao*, 15 November 1922). Fourteen Japanese journalists interviewed Einstein on his arrival. The restaurant in which they dined was Yi Pin Xiang. The journalist was actually the Shanghai correspondent for the Japanese newspaper *Tokyo Nichinichi Shinbun*, [?] Murata (see *Min Guo Ri Bao*, 14 November 1922; *The China Press*, 14 November 1922; and *Tokyo Nichinichi Shinbun*, 15 November 1922).

51. In the afternoon, they visited the Chenghuang Temple and the Yuyuan Garden in the old city of Shanghai and attended a performance of the traditional Kunqu opera at the Xiao-Shi-Jie (Little World) Theater (see *Min Guo Ri Bao*, 14 November 1922).

52. Apparently, the tea was attended by members of the German community of Shanghai (see *The China Press*, 14 November 1922).

53. The Jewish delegation was headed by the Rabbi of Shanghai, Woolf Hirsch (see *Israel's Messenger*, 1 December 1922).

54. The dinner took place at the residence of Yiting Wang (1867–1938), an entrepreneur, socialite, philanthropist, painter, and Buddhist scholar. The German-speaking

Chinese couple were Shi Ying, dean of the Zhejiang School of Law and Political Science, and his wife, Shu Zhang. Their daughter was Huide Ying. The rector of the University of Shanghai was Youren Yu (1878?–1964). Another prominent guest was Junmou Zhang, a former professor at the University of Beijing. For a group portrait taken during the visit, see Illustration 10.

55. The dinner speeches by Youren Yu and Einstein were published in *Min Guo Ri Bao*, 14 November 1922. In his speech, Einstein expressed his admiration for Wang's art and his belief in the future contribution of Chinese youth to science.

56. Lord Richard B. Haldane (1856–1928), former British secretary of state for war and lord chancellor, was a lawyer and philosopher.

57. The Gakushi-Kai, an association of male graduates from imperial Japanese universities. The reception was held at the Y.P.S. Hotel (see *Israel's Messenger*, 1 December 1922).

58. Hantaro Nagaoka (1865–1950), professor of theoretical physics at Tokyo Imperial University, and Toyo Nagaoka (1870–1946). Jun Ishiwara (1881–1947) was a science writer and journalist at the Iwanami Shoten Publishing Co. in Tokyo and former professor of physics at Tohoku University in Sendai. Ayao Kuwaki (1878–1945) was professor of physics at Kyushu Imperial University in Fukuoka City. The German consul was Oskar Trautmann (1877–1950). The German club was the Japanese-German Society (the Club Concordia), established in 1911.

According to diplomatic and press reports, Einstein arrived at 3 p.m. and was greeted by Ishiwara; Keiichi Aichi, professor of physics at Tohoku Imperial University at Sendai; the noted Japanese pacifist Toyohiko Kagawa; and "several others" (see Governor of Hyogo Prefecture to Minister of Diplomacy, 18 November 1922 [JTDRO, Diplomatic R/3.9.4.110.5], and *Japan Times & Mail*, 17 November 1922). Einstein could not be greeted by the German ambassador, Wilhelm Solf (1862–1936), as the ambassador had not yet returned from a brief stay in Germany (see *Grundmann 2004*, p. 229).

59. On his arrival, Einstein told the many reporters "that his present visit is to see the sights of the country and become acquainted with Japanese art and music, particularly the latter." He also stated that "he is glad to come to Japan as he feels he is thereby promoting human brotherhood through the medium of science." Einstein spoke in German to the reporters, and Elsa translated his replies into English (see *Japan Times & Mail*, 18 November 1922). In another interview, he said, "I always had the wish to see the Land of the Rising Sun ever since I read [the author] 'Lafcadio Hearn' and the 'Tales of Old Japan,' by Lord Redesdale. My wish in coming to these shores was prompted by my intention to make fast the ties among the different nations through intellectual bonds; to make the world of science one international community" (*Osaka Mainichi Shinbun*, English Daily Edition, 18 November 1922).

60. The Oriental Hotel.

61. Einstein departed Kobe at 5:30 p.m. from the San-no-miya railroad station, accompanied by Yamamoto, Ishiwara, and Nagaoka. He arrived at Kyoto station at 7:30

p.m. and lodged at the Miyako Hotel in Kyoto (see Governor of Hyogo Prefecture to Minister of Diplomacy, 18 October 1922, and Governor of Kyoto Prefecture to Minister of Diplomacy, 18 November 1922 [JTDRO, Diplomatic R/3.9.4.110.5]).

62. Einstein's car passed the Kamo-Shrine, the Heian-Shrine, and Kyoto Gosho, the former imperial palace (see *Ishiwara 1923*, pp. 18–19).

63. He departed Kyoto station at 9:15 a.m. (see Governor of Kyoto Prefecture to Minister of Diplomacy, 18 November 1922 [JTDRO, Diplomatic R/3.9.4.110.5], and *Tokyo Nichinichi Shinbun*, 19 November 1922). The train passed Lake Biwa, Lake Hamana, and Mount Fuji. Einstein made a stop at Sekigahara, the site of a major battle in Japanese history.

64. The Einsteins' arrival at the Tokyo railroad station was "like welcoming a general returning from a victorious campaign." Tens of thousands of people gathered on the train platform and in the station square to greet the Einsteins, who were prevented from leaving the platform for more than half an hour (see *Yamamoto 1934*). When the train arrived at 7:20 p.m., the large crowd started shouting "Einstein! Einstein!" as soon as they caught a glimpse of him. He left the station amid cries of "Banzai" (*Tokyo Nichinichi Shinbun*, 19 November 1922). There were such crowds at the station "that the police were forced to helplessly tolerate the life threatening throng" (see Wilhelm Solf to German Foreign Ministry, 3 January 1923 [GyBSA, I. HA, Rep. 76 Vc, Sekt. 1, Tit. 11, Teil 5c, Nr. 55, Bl. 157–158]).

65. Tokyo Teikoku Hoteru (the Tokyo Imperial Hotel), which at the time was being redesigned by Frank Lloyd Wright.

66. On behalf of the Imperial Academy, Einstein was greeted by its president, Nobushige Hozumi, and two other members, Gen-yoku Kuwaki, professor of philosophy at the Tokyo Imperial University, and Yasushi Hijikata, professor of English law at the same university and member of the House of Peers. Approximately fifty persons from academia and Kaizo-Sha were waiting on the station platform, yet "barely managed" to greet the Einsteins (see *Kaneko 1981*, vol. 1, p. 36).

67. Siegfried Berliner (1884–1961) was professor of business administration at Tokyo Imperial University. He was one of the Germans who greeted Einstein at the Tokyo railroad station (see *Tokyo Nichinichi Shinbun*, 19 November 1922). His wife was Anna Berliner.

68. Einstein delivered his first public lecture on special and general relativity in the Mita Grand Lecture Hall at Keio University. The audience numbered two thousand, was "composed of men of all walks of life, students and men of science predominating," and included Eikichi Kamada, the minister of education. Elsa Einstein wore a kimono to the lecture, which was greeted with much applause. According to press reports, Einstein tried to render the lecture comprehensible for the general audience; however, it became technical at times in its second half. Einstein spoke without notes and paused at approximately fifteen-minute intervals to allow Ishiwara to translate (see *Japan Times & Mail*, 20 November 1922; *Osaka Mainichi Shinbun*, English Daily Edition, 21 November 1922; "Pressebericht vom 5.

Dezember 1922" [GyBSA, I. HA, Rep. 76 Vc, Sekt. 1, Tit. 11, Teil 5c, Nr. 55, Bl. 147]; and *Ezawa 2005*, p. 9). The admission fee to the public lectures was three yen for adults and two yen for students, which was "equivalent to the cost of ten ordinary lunches" (see *Kaneko 1987*, p. 357).

69. The luncheon was held at the Koishikawa Botanical Gardens in Tokyo. It was hosted by Nobushige Hozumi, and attended by some forty members of the Academy, including Hantaro Nagaoka, Tetsujiro Inoue, Shibasaburo Kitasato, Tokuzo Fukuda, and Minister of Justice Keijiro Okano. Nagaoka drafted the welcoming address (see *Kaneko 1987*, p. 379). For the German version of the welcoming address, see Nobushige Hozumi to Einstein, 20 November 1922 (*CPAE 2012*, Vol. 13, Abs. 452). For the Japanese version, see "Einstein-Sensei Kangei no Ji" (AEA [65 020.1]). For a list of participants at the luncheon, see "Report of the Luncheon Welcome for Professor Einstein" [JTJA].

70. Sanehiko Yamamoto (1885–1952) was the president of the publishing house Kaizo-Sha. In the aftermath of Einstein's tour, the German Government Headquarters for Scientific Reporting sent a report to the Foreign Ministry alleging that Einstein's trip was funded by *Kaizo*, "the Communist newspaper." This prompted the Foreign Ministry to verify the veracity of this report with the German Embassy in Tokyo (see Karl Kerkhoff to Otto Soehring, 11 January 1923 [GyBPAAA, R64677], and Otto Soehring(?) to German Embassy in Tokyo, 27 January 1923 [GyBPAA, R85846]).

71. A performance of Kabuki theater at the Meiji Theater, which was established in 1893 as a popular theater for Kabuki and Shimpa drama.

72. The chrysanthemum-viewing garden party was held in the gardens of the Akasaka Palace, the imperial detached palace in Tokyo. This tradition began in 1880. For the program and entrance passes to the party, see "Programme" and "November 21st 1922" [JTNAJ]. According to the German ambassador, the garden party was "the culmination of the honors" bestowed on Einstein: members of the German Embassy who attended the party described "how, due to Einstein, the approximately 3,000 participants at this traditional festival of the union of royal family with the people completely forgot the significance of the celebration;" see Wilhelm Solf to German Foreign Ministry, 3 January 1923 [GyBSA, I. HA, Rep. 76 Vc, Sekt. 1, Tit.11, Teil 5c, Nr. 55, Bl. 157–158]. Press reports claimed that six hundred people attended the garden party, including such prominent guests as the Japanese prime minister, Viscount Kato Tomosaburo, and other Japanese and foreign statesmen, businessmen, and military officers (see *Hinode Shinbun*, 22 November 1922).

73. Ernst Bärwald (1885–1952) was the representative of the German company I.G. Farben in Tokyo.

74. Empress Sadako Kujo (Teimei) (1884–1951), the empress consort of Emperor Taishō. The emperor himself had retired from public life to his country villas at the end of 1919 and seldom made appearances in Tokyo (see *Seagrave and Seagrave 1999*, p. 81).

75. Inagaki and his wife. Sanehiko Yamamoto.

76. In Asakusa, Tokyo's major entertainment district.

77. The Zojo Temple in Tokyo.

78. The traditional Japanese dishes sukiyaki and sushi were served (see *Inagaki 1923a*, p. 178, and *Tokyo Nichinichi Shinbun*, 23 November 1922).

79. Yoshi Yamamoto and their children, Misako and Sayoko. Einstein also toured the Meiji Shrine, a memorial for the previous emperor (see *Kaneko 1981*, vol. 1, p. 258).

80. President Nobushige Hozumi and and his son-in-law Motoji Shibusawa (1876–1975), who had studied with Heinrich Friedrich Weber (1843–1912), Einstein's former physics professor at the Polytechnic in Zurich.

81. Rikitaro Fujisawa (1861–1933), former professor of mathematics at Tokyo Imperial University. Kenji Sugimoto claims that Einstein mistook Hozumi for Fujisawa, and, due to his falling out with Weber over his dissertation, he was disgruntled about the planned reception (see *Sugimoto 2001b*, p. 33).

82. The German Society for Natural History and Ethnology of East Asia.

83. The Nippon Kogaku Kogyo company, the predecessor of Nikon, had invited eight German specialist workers to their Oimachi factory for their expertise (see *Long 2006*, p. 11).

84. M. H. Schultz was a chief registrar at the German Embassy in Tokyo.

85. Kaizo-Sha invited journalists from various Japanese newspapers for a luncheon party. Einstein voiced his opinions on the use of rickshaws, Japanese hygiene, and the intrusion by the Japanese press into the private lives of individuals (see *Inagaki 1923a*, p. 179).

86. The Tokyo School of Music, Japan's first official music academy, founded in 1887.

87. Ernst Bärwald, and Nagayoshi Nagai (1845–1929), who was professor of chemistry at Tokyo Imperial University.

88. In the Ginza shopping district (see *Inagaki 1923a*, p. 179).

89. Kaichiro Nezu (1860–1940) was a prominent entrepreneur and art collector. Einstein visited the museum with Yoshimori Yazaki, a graduate in philosophy from Tokyo Imperial University (see his account in *Kaizo*, January 1923).

90. Einstein delivered his second popular lecture at the Youth Assembly Hall (Kanda Seinenkaikan) in Tokyo. The title of the lecture was "On Space and Time in Physics." The hall was so congested that "several dozens of people who had entrance tickets . . . were unable to enter." Kaizo-Sha therefore paid for their return fare to Sendai, where the next lecture was to take place (see *Ezawa 2005*, p. 9, and *Kaneko 2005*, p. 13). This second lecture was characterized as being "on a much broader base" than the first in that it explained time and space, rather than the special and general theories of relativity (see *Osaka Mainichi Shinbun*, English Daily Edition, 28 November 1922).

91. Einstein delivered his first scientific lecture at the main auditorium of the Department of Physics at Tokyo Imperial University. According to *Sugimoto 2001a*, pp. 10–11, the title of the first lecture in the series was "Lorentz Transformation, Special

Theory of Relativity." However, according to *Ishiwara 1923*, p. 88, the title was "Special Theory of Relativity." Einstein's scientific lectures were attended by "120 professors and the like, 5 graduate students and 18 undergraduates." An incomplete list of the participants and an account of the lectures appeared in the January issue of *Kaizo* (see *Ezawa 2005*, pp. 8 and 11).

92. The reception was hosted by the entire student body at the university. The event was held at the university's Faculty of Law Octagon Hall. Following an introduction by Hantaro Nagaoka, Tokudo Takeuchi, a third-year student in the Department of Politics, welcomed Einstein on behalf of the student body (see *Tokyo Nichinichi Shinbun*, 26 November 1922).

93. The Ichimura Theater, one of the oldest Kabuki theaters in Japan, was established in the seventeenth century. Following the performance, Einstein visited the dressing room to thank the dancers personally (see *Tokyo Asahi Shinbun*, 26 November 1922).

94. This reception was held by the Metropolitan Press Association at the Hirano-Ya Ryokan, a traditional Japanese inn.

95. Morikatsu Inagaki.

96. The Museum Shoko-Kan (Antique Collecting House), established by Kihachiro Okura (1837–1928), a merchant and art collector.

97. He may have seen the Noh performance at the Hoso Kai Theater (see Einstein to actors of the Hoso Kai Theater, after 25 November 1922 [*CPAE 2012*, Vol. 13, Abs. 457]). For other instances of Einstein's impression of the Noh theater, see Text 4 in the Additional Texts section of this volume and *Kuwaki 1934*.

98. At the Maruzen publishing house, which was in the vicinity of Einstein's hotel (see *Sugimoto 2001b*, p. 45).

99. Tatsuji Okaya, a former student of Nagaoka, and Fumi Okaya (1898–1945). For the menu at the luncheon at Nagaoka's home, see "Déjeuner," 27 November 1922 (NjP-L, Einstein in Japan Collection, box 1, folder 1, C0904). See Illustration 13.

100. Einstein gave his second scientific lecture in the main auditorium of the Department of Physics at Tokyo Imperial University. According to *Sugimoto 2001a*, pp. 10–11, the title of the second lecture was "Four Dimensional Space, Tensor Analysis." However, according to *Ishiwara 1923*, the title was "Special Theory of Relativity," the same as the first lecture.

101. Marquis Yoshichika Tokugawa (1886–1976), a graduate of Tokyo Imperial University, botanist, and head of the Owari branch of the former shogunal family. He had also sailed on the S.S. *Kitano Maru* from Europe (see *Jansen 1989*, p. 152).

102. Christoph W. Gluck; Miksa (Michael) Hauser (1822–1887), Austro-Hungarian violinist and composer; Johann S. Bach; Henryk Wieniawski (1835–1880), Polish violinist and composer. For reminiscences of Einstein's playing, see *Inagaki 1923b*.

103. The Tokyo University of Commerce (the present-day Hitotsubashi University). In his reply to the welcoming address, Einstein expressed his belief that it was through the genre of art that Japan made an important contribution to world culture (see *Nagashima 1923*). For the greetings by the university's student association,

see Student Association, University of Commerce, Tokyo to Einstein, 28 November 1922 (*CPAE 2012*, Vol. 13, Abs. 461).

104. The title of Einstein's speech was "To Japanese Youngsters." For press coverage, see *Osaka Mainichi Shinbun*, English Daily Edition, 30 November 1922 . The rector was Zensaku Sano (1873–1952).

105. Einstein's third scientific lecture was held in the main auditorium of the Department of Physics at Tokyo Imperial University. According to *Sugimoto 2001a*, pp. 10–11, the title of the lecture was "Tensorial Representation of the Space-Time." However, according to *Ishiwara 1923*, the title was "Special Theory of Relativity," the same as the previous two lectures.

106. A Chinese restaurant at the Shinbashi railroad station, attended by some thirty to forty employees of the Kaizo-Sha (see *Inagaki 1923a*, p. 183).

107. Possibly the passage on music in Text 4 in the Additional Texts section of this volume. In the draft to that document, Einstein added the following note, which may be the "short article" he references here: "On music, dictated to Mrs. I.; missing here. E." "Mrs. I." may be the German-speaking wife of Inagaki.

108. Neil Gordon Munro (1863–1942) was a Scottish physician with a passion for Japanese culture and archeology.

109. Einstein apparently asked Yamamoto to arrange for him to attend a tea ceremony. The businessman and tea-master Yoshi (also known as Soan) Takahashi (1861–1937) subsequently invited Einstein to the ceremony. There are conflicting reports about where the ceremony was held. According to *Kaneko 1984*, it was held at the Garando-Ichiki-an in Akasaka, Tokyo. However, according to *Inagaki 1923a*, it was held in a private tea room in Takahashi's house. The multi-volume work was *Takahashi 1921–1927*, which eventually numbered ten volumes (see *Kaneko 1984*, p. 65). For Takahashi's reminiscences of his meeting with Einstein, see *Takahashi 1933*.

110. Marquis Shigenobu Okuma (1838–1922) had served as minister of finance and foreign affairs in the Meiji period and as prime minister in the Taishō period. Waseda University was founded in 1882 and guided by the principle of "independence of learning" (see *Waseda 2010*, p. 8). President Masasada Shiozawa delivered a welcome address, and in his reply, Einstein remarked that he had noticed the unexpected progress of the Japanese scholarly community and that he was looking forward to its future contributions (see *Waseda Gakuho*, 10 January 1923).

111. According to both *Sugimoto 2001a*, pp. 10–11, and *Ishiwara 1923*, the title of Einstein's fourth scientific lecture was "On the General Theory of Relativity."

112. At the Tokyo Women's Higher Normal School (the present-day Ochanomizu Women's University), a teacher training institute. The reception was presided over by the Imperial Pedagogical Association and eleven other educational associations. It was attended by one thousand people (see *Taisho 11 nen Nisshi Tokyo-joshi-koto-shihan-gakkou*, 29 November 1923).

113. Einstein visited the Division of Imperial Court Music at the Imperial Household Agency and attended a performance of *gagaku*, Japanese ancient court music and dance (see *Yomiuri Shinbun*, 1 December 1922, and *Aichi 1923*, p. 300).

114. According to *Sugimoto 2001a*, pp. 10–11, the title of Einstein's fifth scientific lecture was "On the Equation of the Gravitational Field." Yet according to *Ishiwara 1923*, it was titled "General Theory of Relativity." Takuro Tamaru (1872–1932) was professor of physics at Tokyo Imperial University. Uzumi Doi (1895–1945) was a graduate student under Hantaro Nagaoka at Tokyo Imperial University and lecturer in physics at the prestigious First Higher School. For his book challenging the theory of relativity, see *Doi 1922*. According to press reports, Doi, who had criticized Einstein's theories, admitted that he had been wrong and asked Tamaru to read his statement in German to that effect (see *Tokyo Asahi Shinbun*, 2 December 1922, and *Kahoku Shimpo*, 1 December 1922). Earlier, Einstein had confirmed that he had read the pamphlet Doi had sent him to Berlin and that it merited "serious study." However, he was not concerned that it would pose a challenge to the theory of relativity (see *Tokyo Nichinichi Shinbun* and *Osaka Mainichi Shinbun*, 18 November 1922). For Doi's diary during Einstein's visit, in which he stated that he retracted his rejection of his own theory merely half an hour after they met, see EPPA, 95 077. On the controversy between Doi and Keiichi Aichi over Einstein's theories, see *Osaka Mainichi Shinbun,* English Daily Edition, 5 November 1922.

The welcoming message from the students of the university is available (see Students of Tokyo Imperial University to Einstein, 30 November(?) 1922 [*CPAE 2012*, Vol. 13, Abs. 464]).

115. The Danish ambassador was Niels Höst (1869–1953).

116. According to *Sugimoto 2001a*, pp. 10–11, Einstein's sixth (and final) scientific lecture was titled "On the Cosmological Problem." However, according to *Ishiwara 1923*, it was titled "General Theory of Relativity." To mark the end of the lecture series, a commemorative photo of Einstein and Japanese scientists was taken beside Sanshiro Pond at the heart of the university. The photograph was presented to Einstein with an autograph album signed by the faculty and students of the physics department. The album (see NNLBI, Albert Einstein Collection: Addenda [AR 7279]) also included a letter of appreciation written by Hantaro Nagaoka and signed by him and 124 other signatories (see Hantaro Nagaoka et al. to Einstein, 1 December 1922 [*CPAE 2012*, Vol. 13, Doc. 389]). For a description of these events, see *Ishiwara 1923*, pp. 111–112.

117. The banquet was held at the Imperial Hotel in honor of the conclusion of the lecture series. It was attended by 150 scholars, writers, and employees of Kaizo-Sha. Among the attendees were Hantaro Nagaoka, Jun Ishiwara, Ayao Kuwaki, Takeo Arishima, Takuro Tamaru, Tetsujiro Inoue, Torahiko Terada, and Shinzo Koizumi (see *Kahoku Shimpo*, 3 December 1922, and *Kaneko 1981*, vol. 1, p. 259).

118. The Tokyo School of Technology (the present-day Tokyo Institute of Technology), founded in 1881. Tokio Takeuchi (1894–1944) was assistant professor of physics there.

119. Einstein arrived at Sendai station at 9:17 p.m. (see Governor of Miyagi Prefecture to Minister of Diplomacy, 6 December 1922 [JTDRO, Diplomatic R/3.9.4.110.5]).

Kotaro Honda (1870–1954) and Keiichi Aichi (1880–1923) were both professors of physics at Tohoku Imperial University at Sendai. They traveled from Sendai to Kooriyama station, roughly midway between Tokyo and Sendai (see *Kahoku Shimpo*, 4 December 1922).

120. The prominent physicists at Tohoku Imperial University were Shirota Kusakabe, Yoshitoshi Endo, and Mitsuo Yamada. The rector was Masataka Ogawa. Hans Molisch (1856–1937) was an Austrian botanist and a professor of biology. It took Einstein twenty minutes to advance from the station to the Sendai Hotel, due to the large crowds. At the hotel, Einstein was greeted by Yuichiro Chikaraishi, governor of Miyagi Prefecture, Takesaburo Kanomata, mayor of Sendai, and Ogawa (see Governor of Miyagi Prefecture to Minister of Diplomacy, 6 December 1922 [JTDRO, Diplomatic R/3.9.4.110.5], and *Kahoku Shimpo*, 4 December 1922).

121. Einstein's third popular lecture, titled "On the Principle of Relativity," was given at the Sendai Civic Auditorium. It was translated by Keiichi Aichi and apparently held for free to compensate for the audience members who had not been able to enter the lecture at the Kanda Youth Hall in Tokyo. The audience numbered 350 and consisted mainly of professors and university students (see Governor of Miyagi Prefecture to Minister of Diplomacy, 6 December 1922 [JTDRO, Diplomatic R/3.9.4.110.5]). For accounts of the lecture, see *Yomiuri Shinbun*, 4 December 1922, and *Okamoto 1981*, pp. 931–932.

122. Ippei Okamoto (1886–1948) was a painter in the Western tradition and a cartoonist for the *Tokyo Asahi Shinbun* newspaper. He joined Einstein's entourage "on his own, out of personal admiration and the wish to observe the great scientist at close range" (see *Okamoto 1981*, p. 931). During Einstein's tour, he contributed articles to his newspaper (see *Tokyo Asahi Shinbun*, 9–15 December 1922). Following Einstein's departure, he published *Okamoto 1923*.

The Matsushima Islands, a group of some 260 tiny pine-clad islands near Sendai. During the train journey to the islands, Okamoto sketched Einstein, who signed the sketch with "Albert Einstein or The Nose as Thought-Reservoir"; see *Okamoto 1981*, p. 932). See Illustration 18.

123. They dined at the Matsushima Hotel (see *Kaneko 1981*, vol. 2, p. 34) and visited the Zuiganji Temple on Matsushima (see *Okamoto 1981*, p. 933). Bansui Tsuchii (Doi) (1871–1952) was a poet and scholar of English. Katsushika Hokusai (1760–1849) was one of the best-known woodprint artists in the Edo period. Tsuchii offered Einstein two bound sets of woodblocks to choose from, Utagawa Hiroshige's *Fifty-Three Stations on the Tokaido* and Hokusai's *One Hundred Views of Mount Fuji* (see *Okamoto 1981*, p. 932). Okamoto inscribed the flyleaf of the album with a dedication "Albert Einstein, in cordial gratitude," in German, and "at Sendai. Taisho 11, Twelfth month" above the signature "Drawn by Ippei," in Japanese (see *Jansen 1989*, p. 145). For the Italian book of poetry, see *Tsuchii 1920*.

124. At Tohoku University in Sendai. At the student reception, the university's president Masataka Ogawa led the students in a chorus of "Banzai" for Einstein. More

than fifty professors welcomed Einstein in a conference hall at the College of Engineering (see *Okamoto 1981*, p. 933). The dean of the medical faculty was Toshihiko Fujita (1877–1965). Einstein inscribed his name on a wall in a conference room at the university, beneath the signature of Hans Molisch. The inscription reads: "Albert Einstein 3.XII 22." [ JSeTU, 95 037]. There were reports in the press that Einstein was offered a position at Tohoku University as a temporary professor of physics. A salary of ten thousand yen (approximately five thousand U.S. dollars) and a residence were allegedly proposed (see, e.g., *Osaka Mainichi Shinbun*, English Daily Edition, 5 December 1922). Such reports even led to rumors that Einstein planned to emigrate to Japan; they were then denied by the German Foreign Ministry (see Otto Soehring to the German Consulate in Geneva, 9 December 1922 [GyBPAAA/R 64677]).

125. Kotaro Honda.

126. Okamoto's wife was the well-known Japanese novelist and poet Kanoko Okamoto (1889–1939). Elsa Einstein and Tony Inagaki had remained in Tokyo.

127. They arrived at Nikko railroad station at 4:10 p.m. and stayed at the Kanaya Hotel in Nikko (see Governor of Tochigi Prefecture to Minister of Diplomacy, 7 December 1922 [ JTDRO, Diplomatic R/3.9.4.110.5], and the hotel registry with Einstein's signature (AEA [122 789]). Elsa Einstein arrived separately with Tony Inagaki later that day from Tokyo (see *Okamoto 1981*, p. 935). For a list of the ten subjects "drawn . . . on ten-inch squares of finely finished cardboard" and presented to Einstein, see *Okamoto 1981*, p. 937.

128. Tony Inagaki.

129. Einstein, Inagaki, and Okamoto arrived at the Chugu Shrine on Lake Chuzenji in Nikko at 10 a.m. They toured the Hoto, Hannya, and Kegon waterfalls. They returned to the hotel at 4 p.m. (see Governor of Tochigi Prefecture to Minister of Diplomacy, 7 December 1922 [ JTDRO, Diplomatic R/3.9.4.110.5], and *Okamoto 1981*, p. 935).

130. For Okamoto's account of these conversations, see *Okamoto 1981*, pp. 935–936.

131. Einstein, Elsa, and others visited Toshogu Temple and "other related temples" (see Governor of Tochigi Prefecture to Minister of Diplomacy, 7 December 1922 [ JTDRO, Diplomatic R/3.9.4.110.5]).

132. The Tokugawa shogunate ruled Japan during the Edo period from 1603 to 1868. Ieyasu Tokugawa (1543–1616) was its founder and first shogun.

133. They departed Nikko for Tokyo at 5:10 p.m. (see Governor of Tochigi Prefecture to Minister of Diplomacy, 7 December 1922 [ JTDRO, Diplomatic R/3.9.4.110.5]).

134. Ernst Bärwald. For the article, see Text 4 in the Additional Texts section of this volume.

135. Einstein arrived at Nagoya station at 4:41 p.m. He was greeted by executives of the Shin-Aichi company, principals, professors, and approximately one thousand students of the Medical College and higher schools who shouted "Banzai" (see *Shin Aichi*, 8 December 1922).

136. Leonor Michaelis (1875–1949) was a German-born professor of biochemistry at Aichi Medical College.

137. The Atsuta Shrine.

138. A luncheon at the hotel was held by the Kaizo-Sha and the Shin Aichi Co. When the Einsteins departed Nagoya at 4:46 p.m., a large crowd came to bid them farewell (see *Shin Aichi*, 10 December 1922). The Einsteins arrived at Kyoto station at 7:38 p.m. and stayed at the Miyako Hotel (see Governor of Kyoto Prefecture to Minister of Diplomacy, 11 December 1922 [JTDRO, Diplomatic R/3.9.4.110.5]). The Chion-in Temple in Kyoto, at which "the great bell, not normally struck for anyone, is struck" for Einstein (see *Okamoto 1981*, p. 937).

139. At this point in the original text, Einstein appended the note in the right margin: "8th and 9th in the wrong sequence."

140. Nagoya Castle.

141. Einstein's fourth popular lecture, titled "On the Principle of Relativity," was held at the Gymnasium for National Sport (Nagoya Kokugikan) and was translated by Jun Ishiwara (see *Ishiwara 1923*).

142. At this point in the original text, Einstein appended a note at the right margin: "wrong sequence."

143. Einstein departed (without Elsa) from Kyoto station at 10:40 a.m. and arrived at Osaka station at 11:32 a.m., accompanied by German ambassador Solf, Ishiwara, and Yamamoto. Einstein and Solf attended a welcoming reception held by the Japanese-German Society at the Osaka Hotel. The reception was attended by two hundred people. In his reply to Sata's welcome, Einstein stressed that "he took it that the enthusiastic welcome was not intended for him only, but for German science as a whole, and, only in that spirit, could he accept it." The military band of the Osaka garrison played the Japanese and German national anthems, and the reception ended with "Banzais" for both nations (see Governor of Kyoto Prefecture to Minister of Diplomacy, 11 December 1922, and Governor of Osaka Prefecture to Minister of Diplomacy, 14 December 1922 [JTDRO, Diplomatic R/3.9.4.110.5], and *Osaka Mainichi Shinbun*, English Daily Edition, 12 December 1922). The mayor of Osaka was Shiro Ikegami (1857–1929). Aihiko Sata (1871–1950) was professor of pathology, president of Osaka Medical College, and president of the Japanese-German Society.

144. Einstein's sixth popular lecture, titled "On the General and Special Principles of Relativity," was held at Osaka Central Auditorium at 6 p.m. and translated by Ishiwara. It was attended by two thousand people. He returned to Kyoto the same day by departing Osaka at 10:22 p.m. (see Governor of Osaka Prefecture to Minister of Diplomacy, 14 December 1922 [JTDRO, Diplomatic R/3.9.4.110.5], and *Ezawa 2005*, p. 9).

145. Einstein's fifth popular lecture, titled "On the Principle of Relativity" was held at Kyoto Civic Auditorium and was interpreted by Ishiwara (see *Osaka Mainichi Shinbun*, English Daily Edition, 8 December 1922).

146. The Sento Imperial Palace in Kyoto. For Ishiwara's description of Einstein's visit to the palace, see *Ishiwara 1923*, pp. 155–157.

147. The portraits of thirty-two Chinese sages are depicted on eight sliding doors made of paper. They originate in the Heian period (between 794 and 1185).

148. The shrine to Robert Koch (1843–1910) was established by one of his dedicated Japanese students, Shibasaburo Kitasato, intially at the National Institute for the Study of Infectious Diseases and then moved to the Kitasato Institute, both in Tokyo.

149. The Nijo castle in Kyoto, established by Ieyasu Tokugawa.

150. For details on their joint work, see Jun Ishiwara to Einstein, 12 January 1923 [*CPAE 2012*, Vol. 13, Doc. 422] and Jun Ishiwara to Einstein, after 26 February 1923 or after 21 March 1923 [*CPAE 2012*, Vol. 13, Doc. 433].

151. Toyohiko Kagawa (1898–1960) was a Christian reformer and labor activist. Blank left in original for his name. For Kagawa's impressions of his two meetings with Einstein, see *Kaneko 1987*, p. 369.

152. Einstein's seventh popular lecture, titled "On the Principle of Relativity," was held at the Kobe YMCA and translated by Ishiwara. The reception at the German Club was held under the auspices of Oskar Trautmann, the German consul-general (see *Osaka Mainichi Shinbun*, English Daily Edition, 15 December 1922).

153. The Kyoto Imperial University. The rector was Torasaburo Araki (1866–1942). The student representative was Toshima Araki (1897–1978). For his greetings to Einstein on behalf of the student association, see Toshima Araki to Einstein, 10 December 1922 [*CPAE 2012*, Vol. 13, Abs. 467].

154. An impromptu lecture, titled "How Did I Create the Theory of Relativity?" and initiated by Kitaro Nishida, was held at the main auditorium of the Law Faculty at Kyoto University and translated by Ishiwara (see *Osaka Asahi Shinbun*, 15 December 1922, and *Ezawa 2005*, p. 10). For Ishiwara's transcription of the lecture, see "How I Created the Theory of Relativity," 14 December 1922 [*CPAE 2012*, Vol. 13, Doc. 399].

155. Masamichi Kimura (1883–1962) was professor of physics at Kyoto Imperial University.

156. One of the presents was a traditional *naga juban* (underkimono) for Elsa (see *Nakamoto 1998*, p. 77).

157. The Chion-in Temple in Kyoto. For a photo of Einstein on the steps of the temple, see Illustration 20.

158. Possibly the Yasaka Shrine and the Shijo Street shopping district, both in close proximity to the Miyako Hotel.

159. The Shogunzuka Dainichido Temple.

160. The Nishi Honganji Temple is the main temple of the Jôdo Shinshu Honganji branch of the Buddhist True Pure Land sect located in Kyoto.

161. Lake Biwa, located northeast of Kyoto, is the largest lake in Japan. Mii Temple is one of the oldest temples in Japan.

162. This is most likely a reference to the world-famous Nishijin brocade.

163. Nara lies 480 kilometers south of Kyoto. The Nara Hotel, where Einstein played the piano (see *Sugimoto 2001b*, p. 112)

164. The most prominent of these temples is the Grand Shrine of Kasuga, founded in 768. The Todaiji Temple houses the large Buddha figure, known as Rushana-butsu zazo and constructed in 745–752.

165. The Nara National Museum. Einstein also visited Nara Park (see *Sugimoto 2001b*, p. 114).

166. Mount Wakakusa.

167. Miyajima Island is in the Aki district of Hiroshima.

168. The Itsukushima Shrine.

169. The holy Mount Misen. The Seto Inland Sea.

170. The German ambassador to Tokyo, Wilhelm Solf. Solf reported that his personal relations with Einstein "developed into friendly ones." Regarding the dispatch, Solf informed Berlin that the *Japan Advertiser* had published a report from the Kokusai-Reuter news agency, according to which the German-Jewish journalist and critic Maximilian Harden had testified in court in Berlin at the trial of his would-be assassins that "Professor Einstein went to Japan because he did not consider himself safe in Germany." Harden's actual quote was "So what has been achieved? The great scholar Albert Einstein is now in Japan because he does not feel safe in Germany." As Solf feared that this report may adversely impact "the extraordinarily beneficial effect of Einstein's visit for the German cause," he requested that Einstein allow him to deny it by cable (see Wilhelm Solf to German Foreign Ministry, 3 January 1923 [GyBSA, I. HA, Rep. 76 Vc, Sekt. 1, Tit. 11, Teil 5c, Nr. 55, Bl. 157–158], and *Neumann and Neumann 2003*, p. 187). For Einstein's subsequent response to Solf, see Text 7 in the Additional Texts section of this volume.

171. Moji is a city in the northern Fukuoka Prefecture, Kyushu. The Einsteins departed Miyajima at 4:10 p.m. and arrived at Shimonoseki at 8:50 p.m. They transferred to a ferryboat to travel to Moji, which lies across the Kanmon Straits. At 9:30 p.m., they arrived at Moji, where they were greeted by Sonta Nagai, the branch manager of Mitsui Bank. Allegedly, Einstein claimed in the interview that "this lifestyle of the Japanese adapting to nature is unlimitedly precious. If possible, I would like to enjoy this Japanese life and style forever. If the circumstances permit it, I even want to live in Japan hereafter" (see Governor of Fukuoka Prefecture to Minister of Diplomacy, 6 January 1923 [JTDRO, Diplomatic R/3.9.4.110.5]; *Fukuoka Nichinichi Shinbun*, 25 December 1922; and *Nakamoto 1998*, pp. 45–46). According to another press report, Einstein remarked that the lack of a truly democratic electoral system in Japan was a serious impediment to the country's development (see *Yomiuri Shinbun*, 25 December 1922). The Mitsui Club was a social club established by the Mitsui-Bussan Co. in 1921.

172. Einstein arrived at Hakata station in Fukuoka at 12:41 p.m. His eighth popular lecture, titled "On the Special and General Principles of Relativity," was held at the Hakata Daihaku Theater in Fukuoka, translated by Ishiwara, and attended by

more than three thousand people (see Governor of Fukuoka Prefecture to Minister of Diplomacy, 6 January 1923 [JTDRO, Diplomatic R/3.9.4.110.5], and *Ezawa 2005*, p. 9). On the lecture in Sendai, see note 90.

173. The banquet organized by the Kaizo-Sha was held at the Café Paulista in the Hakata district. The meeting taking place in the adjacent room was of the alumni association of the Kyushu School of Physics (see *Fukuoka Nichinichi Shinbun*, 26 December 1922).

174. Hayari Miyake. Einstein stayed at the Sakayeya-Ryokan Hotel in the Hakata district. The landlady of the hotel was Tatsu Kuranari (see *Fukuoka Nichinichi Shinbun*, 25 December 1922, and *Nakamoto 1998*, p. 61).

175. On one of the banners, Einstein wrote "Sakayeya A. Einstein. 1922."

176. They returned to the Sakayeya Hotel.

177. Weil's disease is a severe form of leptospirosis, a bacterial infection. The festive banquet was held at Kyushu Imperial University (see Governor of Fukuoka Prefecture to Minister of Diplomacy, 6 January 1923 [JTDRO, Diplomatic R/3.9.4.110.5]). Its president was Bunji Mano (1861–1946).

178. For a press report on the visit to the institute, see *Fukuoka Nichinichi Shinbun*, 26 December 1922.

179. Hayari Miyake. During his visit, Einstein played on Miyake's grand piano, which had recently arrived from Germany (see *Hiki 2009*, pp. 39–40). The Prefectural Showcase House was located at the Prefectural Office. The provincial governor was Ushimaro Sawada.

180. Einstein departed Hakata station at 4:03 p.m. for Moji. At the children's Christmas party held at the Moji YMCA, he played "Ave Maria," accompanied on the piano by Chiyoko Ishikawa, a music teacher at the Shimonoseki Girls High School (see *Fukuoka Nichinichi Shinbun*, 27 December 1922). He spent the night back at the Mitsui Club (see Governor of Fukuoka Prefecture to Minister of Diplomacy, 6 January 1923 [JTDRO, Diplomatic R/3.9.4.110.5]).

181. Mount Otani. Possibly an erroneous reference to the preface to the Japanese edition of his works, which is dated "27. XII. 1922," i.e., the next day (see *Einstein 1923c*).

182. The boat trip was in the Straits of Kanmon. Watanabe was a counselor for the Moji branch of the Mitsui-Bussan Co. Sonta Nagai. For the poem and the drawing, see *Ishiwara 1923*, on verso of frontispiece.

183. The Chamber of Commerce and Industry.

184. According to a press report, on the way to the port in Moji, the Einsteins saw an individual at the roadside pounding rice cakes and shouting in celebration of the New Year. They stopped out of curiosity. Einstein allegedly donned a red headband and joined in the pounding and shouting (see *Tokyo Asahi Shinbun*, 30 December 1922).

185. The S.S. *Haruna Maru* was owned by the Nippon Yusen Kaisha shipping company. It was built in 1921 and sailed the route from Yokohama to Antwerp. In 1942, it ran aground and sank near Omaesaki, Japan. Ayao and Tsutomu Kuwaki (1913–2000).

Hayari Miyake. Watanabe and Sonta Nagai from the Mitsui-Bussan Co. (see *Fukuoka Nichinichi Shinbun*, 30 December 1922).

186. Poem by Bansui Tsuchii, "To the Great Einstein" (see Bansui Tsuchii (Doi) to Einstein, before 30 December 1922 [*CPAE 2012*, Vol. 13, Abs. 486]).

187. The S.S. *Haruna Maru* departed at 3 p.m. from Moji port (see Governor of Fukuoka Prefecture to Minister of Diplomacy, 6 January 1923 [JTDRO, Diplomatic R/3.9.4.110.5]). In his farewell message to Japan, Einstein expressed his gratitude for the welcome he had received and stated that he was most deeply impressed by the realization that "there exists a nation which still preserves the graceful artistic tradition and such beauty of minds with such simplicity" (see *Fukuoka Nichinichi Shinbun*, [29?] December 1922).

188. For Einstein's mention of his previous letter, see Einstein to Jun Ishiwara, after 26 February 1923 or after 21 March 1923 [*CPAE 2012*, Vol. 13, Doc. 433].

189. A reference to the theories of Hermann Weyl (*Weyl 1918*) and Arthur Stanley Eddington (*Eddington 1920*).

190. See Einstein to Sanehiko Yamamoto, 30 December 1922 [*CPAE 2012*, Vol. 13, Doc. 413], and Texts 9 and 11 in the Additional Texts section of this volume. Anna Berliner.

191. R. de Jonge, an engineer. The Einsteins lodged at the home of S. Gatton at 9 Dumer Road (see *Min Guo Ri Bao*, 28 December 1922, and *The China Press*, 30 December 1922).

192. The reception was organized by the Shanghai Jewish Communal Association (see *The China Press*, 31 December 1922). For Einstein's speech, see Text 12 in the Additional Texts section of this volume. Speeches were also delivered by Rabbi W. Hirsch and D. M. David, the president of the Jewish Communal Association (see *The China Press*, 3 January 1923).

193. The discussion on relativity was held at the Shanghai Municipal Committee at 6 p.m. by invitation only. It was hosted by the Young Men's Hebrew Association and the Quest Society. It was presided over by its chairman, Herbert Chatley (1885–1955), a civil engineer, and he was assisted by Rabbi Hirsch and de Jonge (who acted as interpreter). The event was held in the form of a question-and-answer session. Notable questions were asked about the Michelson-Morley experiment, the recent eclipse expedition to Australia, and the obscuration of Jupiter's satellites. It was attended by three to four hundred Westerners and only four or five Chinese, among them Junmou Zhang, who asked Einstein about the "psychic research" of Oliver Lodge, which Einstein dismissed as "not serious" (see *Min Guo Ri Bao*, 28 December 1922 and 3 January 1923, and *The China Press*, 30 and 31 December 1922 and 3 January 1923).

194. See *Eddington 1921*.

195. Gobin was apparently one of the two businessmen Einstein met during his previous visit to Hong Kong (see note 41). The French consul-general was Ulysse-Raphaël Réau (1872–1928).

196. See "On the General Theory of Relativity," ca. 9 January 1923 [*CPAE 2012*, Vol. 13, Doc. 417].

197. See Texts 13 and 14 in the Additional Texts section in this volume. An envelope to Planck is available [AEA, 2 096].

198. Alfred Montor. Joan Voûte (1879–1963) was *Assistent* at the Royal Meteorological and Magnetic Observatory in Batavia (present-day Jakarta). Einstein presumably telegraphed and wrote to Voûte to inform him that he would not be arriving in Java, contrary to his original plan (see Einstein to Paul Ehrenfest, 18 May 1922 [*CPAE 2012*, Vol. 13, Doc. 193]). As of early December, he had still apparently intended to travel to Java: Elsa Einstein told a relative that they would be embarking on the voyage to Java on 26 December (see Elsa Einstein to Jenny Einstein, 9 December 1922 [AEA, 75 226]).

199. Abraham Frankel, a Jewish businessman in Singapore, and his wife, Rosa. Their estate was called "Siglap" (see *Ginsburg 2014*, p. 24).

200. Manasseh Meyer and his daughter, Mozelle Nissim. Even though Einstein was to be involved in additional fundraising for the Hebrew University on his return trip to Singapore, the Singapore Zionist Society decided to scale back the plans. They informed the Zionist Organisation in London that "it was not thought advisable to approach the community for further donations for the University, in view of the recent collections, and also in view of the further appeal that will be made next week by Mrs. Caroline Greenfield for her Hospital work. (Hadasah)" (see C. R. Ginsburg to Israel Cohen, 12 January 1923 [IsJCZA, Z4/2685]).

201. Negombo lies 37 kilometers north of Colombo.

202. Carl Hagenbeck (1844–1913) was a dealer in wild animals who founded a private zoo in Hamburg.

203. See *Einstein 1923b*.

204. Troops from France and Belgium had marched into the Ruhr district on 11 January 1923. The immediate reason for the occupation was to secure reparations deliveries of coke and coal from Germany. However, the wider context for the troop movement was the collapse of French-German negotiations on a reparations schedule, and France's frustration with its wartime allies regarding support of its positions vis-à-vis Germany (see *Fischer 2003*, p. 1).

205. The town Al Qantarah El Sharqiyya (Kantara) is in northeastern Egypt on the eastern side of the Suez Canal.

206. At this point, Einstein added "2." in the right margin to indicate 2 February.

207. The railroad route led from Kantara through the Sinai Peninsula to Rafah, Gaza, and Lod (Lydda). Einstein was greeted at the station in Lod by Menachem Ussishkin (1863–1941), president of the Zionist Executive; Ben-Zion Mossinson (1878–1942), member of the General Zionist Council and director of the "Herzliya" Gymnasium; Colonel Frederick H. Kisch, director of the Political Department of the Zionist Executive; Jacob Thon, Director of the Palestine Land Development Company; David Yellin, president of the Va'ad Leumi, the Jewish National

Council; Joseph Meyuchas, president of the Council of Jerusalem Jews; and Meir Dizengoff, mayor of Tel Aviv. In his diary, Kisch gave the following description of Einstein's arrival: "At Lydda dashed across the platform to greet Prof. Albert Einstein. Found him rather tired as he had sat up all night, but I learned later that this was his own fault, as he had insisted on travelling second-class in spite of every effort to persuade him to go into a wagon-lit which had been reserved for him" (see *Ha'aretz*, 4 February 1923, *Jüdische Presszentrale Zürich*, 9 February 1923, and *Kisch 1938*, p. 29).

208. The stations passed on the way from Lod to Jerusalem were Ramleh, Dayr Aban, and Battir.

209. Solomon Ginzberg (1889–1968), inspector of education for the British Mandatory Authority. Einstein had first met Ginzberg during his tour of the United States in 1921, when Ginzberg had served as his secretary (see Einstein to Judah L. Magnes, 18 April 1921 [*CPAE 2009*, Vol. 12, Doc. 122]).

210. Einstein was accompanied by Captain L. G. A. Cust (1896–1962), the aide-de-camp of Sir Herbert Samuel, the British high commissioner (see *The Palestine Weekly*, 9 February 1923). The Einsteins lodged at the official residence of the high commissioner, Government House, in the Augusta Viktoria complex on the Mount of Olives.

Sir Herbert Samuel, 1st Viscount Samuel (1870–1963), British high commissioner for Palestine. Edwin Samuel (1898–1978), member of the headquarters staff of Sir Ronald Storrs, governor of Jerusalem. Hadassah Samuel-Grasovsky (1897–1986) and David Samuel (1922–2014). Herbert Samuel mentions Einstein's stay at Government House in his memoir (see *Samuel 1945*, pp. 174–175).

211. The Old City of Jerusalem.

212. The Dome of the Rock (the Masjid Qubbat As-Sakhrah) on the Temple Mount.

213. The Al-Aqsa Mosque (the Masjid al-Aqsa).

214. The Western Wall (ha-Kotel ha-Ma'aravi).

215. The ramparts of the Old City.

216. Arthur Ruppin (1876–1943) was director of the Palestine Office in Jaffa. At the time of Einstein's visit, Ruppin was raising funds in the United States for a mortgage bank and other Zionist financial institutions (see *Wasserstein 1977*, p. 272, note 3). His wife was Hannah Ruppin-Hacohen (1892–1985).

217. The Bukharian quarter was established by Jews from Bukhara in Central Asia in 1891.

218. Hugo Bergmann (1883–1975) was director of the National Jewish Library, which was established in 1892. Einstein had first met Bergmann during his sojourn in Prague in 1911–1912 (see *Bergman 1974*, p. 390). Bergmann had solicited Einstein's support for the establishment of the Hebrew University in 1919 (see Hugo Bergmann to Einstein, 22 October 1919 [*CPAE 2004*, Vol. 9, Doc. 147]).

219. Hadassah Samuel-Grasovsky.

220. Jericho lies approximately 45 kilometers northeast of Jerusalem. Most likely the Allenby Bridge.

221. This is a reference to Sir Wyndham Deedes (1883–1956), chief secretary of the British Mandatory Authority in Palestine.

222. Einstein visited the garden suburbs of Beth Hakerem in the west of Jerusalem and Talpiot in the south. He visited Beth Hakerem accompanied by Hadassah Samuel, Hannah Ruppin-Hacohen, and Solomon Ginzberg. He toured the new street in the neighborhood, Hechalutz Street. Both neighborhoods were established in 1922 according to plans by the prominent German-Jewish architect Richard Kaufmann (1887–1953) (see *Ha'aretz* and *Do'ar Hayom*, 7 February 1923, and *Kark and Oren-Nordheim 2001*, p. 169).

  Earlier in the day, Einstein had visited the Jerusalem headquarters of the Zionist Executive. According to Frederick H. Kisch, during his visit Einstein "made a little speech explaining the nature of his brain, which he said was such that he was afraid it would be unproductive work for him to attempt to learn Hebrew" (see *Kisch 1938*, p. 30). In the afternoon, he toured the Zionist Organisation's Museum of Agriculture, accompanied by Ussishkin and Tsadok van Friesland, treasurer of the Zionist Executive. In the early evening, a tea party was held at Ussishkin's home, which was attended by the Jewish dignitaries of Jerusalem, senior British officials, and the department heads of the Zionist Executive, among them Norman Bentwich (1883–1971), attorney-general of the mandate; Albert Hyamson, an Anglo-Jewish historian; and Judah L. Magnes (see *Do'ar Hayom*, 7 February 1923; *Ha'aretz*, 7 and 8 February 1923; and *The New Palestine*, 16 February 1923).

223. The National Jewish Library was located at "Beth Ne'eman" on Ethiopia Street. The press reported that Einstein was welcomed by David Yellin and Yeshayahu Press on behalf of the library's board and by its director Hugo Bergmann and his staff. Einstein toured the reading room, and the readers stood up in honor of the guest. The library prepared an exhibit of Hebrew books on mathematics from the beginnings of the Hebrew press, and Einstein was impressed with the beauty of the printing. He requested information on the condition of the library and promised to influence his colleagues in Europe to help raise the necessary funds to transfer to Jerusalem the numerous books that had been collected for the library overseas (see *Do'ar Hayom*, 7 February 1923, and *Ha'aretz*, 8 February 1923).

224. This was Pessach Hebroni (Hevroni) (1888–1963), a teacher at the Hebrew Teachers' Seminar in Jerusalem.

225. Possibly L. G. A. Cust.

226. The "Bezalel" Art Academy was founded in 1906 by Lithuanuan-Jewish artist and sculptor Boris Schatz. Accompanied by Ginzberg, Einstein toured the academy and viewed its permanent exhibit. The academy's deputy director, Ze'ev Raban, showed the guests his new works, and Schatz spoke about the institution's history and presented Elsa with an amulet. Einstein promised to send Schatz the portrait of him by the Jewish artist Emil Orlik for the planned national museum (see *Do'ar Hayom*, 8 February 1923).

227. The official welcome by the Jewish community of Jerusalem was held at the Lämel School (under the joint auspices of the Zionist Executive and the Va'ad Leumi

[National Council]). The press reported that the entire student body from the Jewish schools in Jerusalem lined the street leading to the Lämel School, with each school displaying its flag. On Einstein's arrival at the reception, he was greeted by a loud cheer and the crowd attempted to rush the gates. Einstein was accompanied by Ussishkin; Yellin; Haim Ariav, secretary-general of the Palestine Zionist Executive; and Shmuel Czernowitz, secretary-general of the Va'ad Leumi. A brass band from the Tachkemoni School played Hebrew songs, and a bouquet of flowers was presented to Elsa. The school's auditorium was decorated for the occasion and approximately two hundred participants attended. Ussishkin and Yellin welcomed Einstein on behalf of their respective institutions, expressing their desire to see Einstein settle in Palestine. Yellin presented Einstein with a scroll that was inscribed with the names of the heads of the various Jewish institutions. He was also inscribed in the "Golden Book" of the Jewish National Fund (see *Do'ar Hayom*, 8 February 1923; *The Palestine Weekly* and *The New Palestine*, 9 February 1923; *Ha'aretz*, 11 February 1923). For the Golden Book inscription, see Illustration 23.

The reception was preceded by an official luncheon at Government House attended by the archeologist and architect Ernest T. Richmond, head of the Political Department of the Secretariat, and his wife, Margaret Richmond-Lubbock; a Mrs. Solomon (possibly Flora Solomon, the wife of Harold Solomon, the controller of stores in the British Mandate); and several Catholic monks: Gaudenzio Orfali, a Franciscan archeologist; Antonin-Gilbert Sertillanges, a Dominican philosopher; Fr. Bertrand Carrière, a Dominican geographer and linguist at the École Biblique in Jerusalem; and Édouard-Paul Dhorme, a Dominican professor of Assyriology at the École Biblique (see *The Palestine Weekly*, 9 February 1923). For a photograph on the occasion of the luncheon, see Illustration 22.

228. Norman Bentwich. Kisch described the dinner as follows: "Dinner at Bentwich's for the Einsteins: a very pleasant party. After dinner some good music by a string quintet in which Einstein played second violin, showing considerable talent and evidently enjoying himself" (see *Kisch 1938*, p. 30).

229. Kisch records in his diary that he accompanied Einstein on a tour of the Old City on the morning of 7 February. On the walk, he explained to Einstein "the political situation and some of the intricacies of the Arab question." For his part, "Einstein spoke of Ussishkin's attempt to persuade him to settle in Jerusalem. He has no intention of doing so, not because it would sever him from his work and friends, but because in Europe he is free and here he would always be a prisoner. He is not prepared to be merely an ornament in Jerusalem" (see *Kisch 1938*, p. 30).

230. The lecture was held in the auditorium of the British Mandate's police school at Gray Hill House on Mount Scopus. The hall was decorated with blue and white stripes and the Union Jack, symbols of the twelve tribes of Israel, the slogan "Light and Learning" ("*ora ve-tora*"), and portraits of Theodor Herzl and Herbert Samuel. The lecture was organized by the Zionist Executive. Invitees included the senior officials of the mandate, Arab dignitaries, the heads of the Christian and Muslim communities, Jewish dignitaries and the heads of the Zionist institutions

in Jerusalem, foreign consuls, members of the scientific community from Jerusalem and Tel Aviv, writers, teachers, and journalists. However, the Arab dignitaries did not attend. The talk was described by the press as "the first scientific lecture held in the temporary halls of the university." The local press described the event in glowing terms: *Ha'aretz* termed the occasion a "national festival and a scientific festival." *Do'ar Hayom* declared that the "Hebrew college" had been opened. Furthermore, the vision of Eliezer Ben Yehuda, the catalyst for the revival of the Hebrew language, was being realized, as Einstein commenced his lecture in Hebrew. Prominent attendees were Sir Herbert Samuel, Sir Ronald Storrs, Chief Rabbi Abraham Isaac Kook, the Hebrew writer and Zionist thinker Ahad Ha'am, Tel Aviv's mayor Meir Dizengoff, and Ben-Zion Mossinson. Menachem Ussishkin greeted Einstein and called on him to "mount the platform which has been waiting for you for two thousand years!"

At the beginning of his lecture, Einstein stated that he was "happy to be reading a lecture in a country which sent out light to the world, and from a house which would send out light to the nations." He regretted he was unable to give his lecture in the language of his nation. In order for his audience to understand his lecture, it was delivered in French and lasted one and a half hours. In his talk, Einstein gave an outline of the theory of relativity, explaining its implications for the understanding of time, space, and gravity. Following the talk, Herbert Samuel expressed his gratitude to Einstein and commented on the significance of his visit to Palestine (see *Do'ar Hayom* and *The Palestine Weekly*, 9 February 1923, and *Ha'aretz*, 11 February 1923).

231. The dinner was hosted at Government House by Herbert Samuel and was attended by Thomas Haycraft, chief justice of Palestine, and Lady Haycraft; Harry Luke, assistant governor of Jerusalem, and Mrs. Luke; Humphrey Bowman, head of the mandate's Department of Education, and Mrs. Bowman; Edward Keith-Roach, first assistant secretary, and Philippa Keith-Roach; Raghib al-Nashashibi, the Arab mayor of Jerusalem, and his wife; W. J. Phythian-Adams of the Palestine Exploration Fund; William F. Albright, American archeologist, and his wife, Ruth Norton; and Hilda Ridler, head of female education in Palestine (see *Ha'aretz*, 8 February 1923, and *The Palestine Weekly,* 9 February 1923).

232. The Einsteins departed Jerusalem from Government House. They arrived in Tel Aviv accompanied by Ben Zion Mossinson and Solomon Ginzberg (see *Ha'aretz*, 8 February 1923, and *The Palestine Weekly*, 9 February 1923).

The reception was held at the Herzliya Gymnasium, which had been constructed in 1909 as the first Hebrew high school in Palestine, in Ahuzat Bayit, the first neighborhood of Tel Aviv. Mossinson introduced Einstein to Ahad Ha'am, the school's board, and its faculty. In his brief speech, Einstein told his audience that he had never seen such a large gathering of Jews. He also expressed his deep admiration for the country's achievements. The guests inspected the building and various classrooms and the students performed gymnastic exercises (see *Ha'aretz*,

9 February 1923; *Do'ar Hayom*, 9 and 12 February 1923; and *The Palestine Weekly*, 16 February 1923).

233. The streets leading to City Hall in Tel Aviv "were lined with throngs of people." On his arrival, Einstein was greeted with applause and by students displaying their schools' flags. The gymnasium's orchestra played for the guests. Mayor Meir Dizengoff and the municipal council members welcomed Einstein and his entourage. Einstein was presented with an address in which he was elected "Honorary Citizen of Tel Aviv" (see Mayor of Tel Aviv [Meir Dizengoff] to Einstein, 8 February 1923 [*CPAE 2012*, Vol. 13, Abs. 514]). This was the first time that such an honor had been bestowed on a visitor. For Einstein's response to this honor, see *Do'ar Hayom*, 12 February 1923. Mossinson told the crowd that Einstein had asked him to inform them that he "was deeply sorry that he could not yet address them in Hebrew but that he was studying the language and hoped to teach you soon in Hebrew at the Jerusalem university." This was greeted by the crowd with cheers and cries of "Long live Prof. Einstein" (see *Ha'aretz* and *Do'ar Hayom*, 9 February 1923, and *The Palestine Weekly*, 16 February 1923). For a photo taken outside the municipality during Einstein's visit, see Illustration 24.

234. The first power station in Tel Aviv was being constructed by Pinhas Ruthenberg, founder of the Palestine Electric Corporation. Tel Aviv's first subterranean electric cable was laid under Allenby Street during Einstein's visit (see *The New Palestine*, 9 February 1923). The quarantine station for immigrants with contagious diseases was located at the port of Jaffa. Einstein toured the Silikat brick factory accompanied by Mayor Dizengoff (see *Ha'aretz*, 11 February 1923). At the time of Einstein's visit, the factory was experiencing a severe labor dispute that had erupted in spring 1922 (see *Shachori 1990*, p. 270).

235. The popular reception was held in the courtyard of the gymnasium. According to press reports, "thousands" attended the event. In his introduction to Einstein's speech, Mossinson claimed that Einstein "has come as a Zionist, to see the country [. . .] in the hope that he will subsequently be able to settle in it." This was greeted by the crowd with cheers. For Einstein's speech, see *Ha'aretz*, 11 February 1923, and *The Palestine Weekly*, 16 February 1923.

236. The agricultural experimental station had been founded by Yitzhak Elazari-Vulkani and was located in the vicinity of the Herzliya Gymnasium. Einstein visited the facility accompanied by Mossinson and Dizengoff (see *Ha'aretz*, 11 February 1923). The courses were held at the Association for Scientific Education on Yehuda Halevi Street. Aharon Czerniawski (1887–1966) was a teacher at the Herzliya Gymnasium. Einstein participated in a meeting of the Association of Engineers and Architects in Palestine, which was held in his honor in Lilienblum Street. Its chairman, Engineer Shimon(?) Reich, presented Einstein with the honorary diploma. The diploma names Einstein as the association's first "honorary member" (see *Ha'aretz*, 11 February 1923, and for the diploma, see The Association of Engineers and Architects in Palestine, to Einstein, 4 February 1923 [*CPAE 2012*, Vol. 13, Abs. 505]).

237. Shmuel Tolkowski (1886–1965) was a citrus grower and member of the Tel Aviv municipal council. The assembly was held in the auditorium of the gymnasium. Only a small number of guests were invited due to the size of the auditorium—among others, public figures, teachers, and writers. Mayor Dizengoff introduced Einstein, yet stated that this was not an easy task, as he had traveled to Jerusalem to hear him lecture and was not embarrassed to admit that he had not understood his lecture. He therefore could not explain to the audience what made Einstein great. Einstein lectured in German on the relationship of the theory of relativity to philosophical issues, e.g., the consequences of relativity for epistemology, its contradiction of Kant's ideas of space and time, and its implications for the finiteness of cosmic space. Following the lecture, the choir of Hanina Karchevsky sang, and the school's orchestra played for the guests (see *Ha'aretz*, 11 February 1923; *Do'ar Hayom*, 11 and 12 February 1923; and *Palestine Weekly*, 16 February 1923). Einstein's diary does not mention two additional locations he toured in the Tel Aviv area: Ir Ganim, a garden city on the outskirts of Tel Aviv (now Ramat Gan), and the public baths located near the Casino coffeehouse by the Mediterranean (see *Do'ar Hayom*, 11 February 1923).

238. Einstein attended the second session of the semiannual conference of the General Federation of Labor at the Eden Cinema in Tel Aviv. Einstein and Elsa entered the hall during the speech of David Ben-Gurion, general secretary of the federation. They were greeted enthusiastically by the delegates. Hugo Bergmann, an executive member of the federation, welcomed Einstein. In his short address to the conference, Einstein stated: "I have observed your work in the country and in particular in Jerusalem with great admiration. I have heard that you are on the path to creating a new Federation of Labor, the likes of which the country has not yet seen; believe me that I am very interested in seeing your work. I do not have much to say to you but much to hear from you; therefore I will be silent." He also stated that he was "convinced that the future of the country and of our people lies in your hands" (see minutes of the second semiannual conference of the Histadrut, March 1923 [AEA, 67-524]; *Ha'aretz*, 11 February 1923; *The Palestine Weekly* and *Jüdische Rundschau*, 16 February 1923).

239. Mikve Israel was founded in 1870 by the Alliance Israélite Universelle. Einstein stopped in Mikve Israel after he had left Tel Aviv by car for Rishon LeZion. He toured the school accompanied by Aharon Czerniawski; Avraham Brill, director of the Palestine Jewish Colonization Association in Judea; and Meir Winik, a chemistry teacher at the school. He visited the school, the dormitory, the nursery, and the dairy and received explanations from the director, Eliyahu Krause. Einstein also visited the grave of the school's founder, Charles Netter, and the vineyards. Following his tour of Mikve Israel, Einstein visited the experimental agricultural farm in Ben Shemen, which had been established by Yitzhak Elazari-Vulkani (see *Ha'aretz*, 11 February 1923). Einstein does not seem to have differentiated between Mikve Israel and the experimental farm.

240. Einstein arrived in the colony of Rishon LeZion accompanied by Czerniawski and Yehuda Nedivi Frankel, an agronomist. He was greeted by horsemen and "almost all the settlers" assembled outside the town hall to welcome him. At the reception held at the community house, Avraham Dov Lubman-Haviv (1864–1951), president of the colony's council, welcomed Einstein on behalf of the municipality. Menashe Meirowitz (1860–1949), president of the agronomists in Palestine, greeted him on behalf of the farmers' association. In his speech, Einstein told his audience that he had seen "energetic people at work, which has impressed me beyond expression." He also promised that "until my last gasp I shall act on behalf of our *Yishuv* [the Jewish community in Palestine] and our land." Following the reception, Einstein visited the colony's schools and was given a tour of the wine cellars by Ze'ev Gluskin, founder of the Carmel wine company. The guests subsequently dined at the Herzliya Hotel (see *Do'ar Hayom*, 11 and 12 February 1923; *The Palestine Weekly*, 16 February 1923).

241. Here and in the next instance, Einstein erred and was actually referring to Haifa.

242. Hillel Jaffe (1864–1936), a physician, Zionist functionary, and board member of the Technion in Haifa, probably the cousin of the Russian physicist Abram F. Ioffe.

The Palestine Salt Company in Atlit, south of Haifa.

243. The Jewish Sabbath begins at sundown on Friday. Hermann Struck (1876–1944) was a German-Jewish artist and one of the founders of the orthodox-Zionist Misrachi party in Germany. He had recently immigrated to Palestine and settled in the Hadar HaCarmel neighborhood in Haifa. Shmuel Yosef Pevzner (1878–1930), Russian Zionist and one of the main builders and developers of the Jewish neighborhoods in Haifa. His wife, Lea Pevzner-Ginzberg (1879–1940), was Solomon Ginzberg's sister. The Einsteins lodged with the Pevzners during their stay in Haifa (see *Do'ar Hayom*, 14 February 1923).

244. Mally (Malka) Struck (1889–1964).

245. According to press reports, Einstein visited the Technion, the technological institute of the Jewish community in Palestine, and not the Reali School, on 10 February. It therefore seems likely that, in recording his impressions of his visits, Einstein switched the days on which he visited the two institutions.

The Hebrew Reali School in Haifa was founded by the Zionist Executive in 1913. The school had recently moved into a former British military hospital building adjacent to the Technion. Arthur Biram (1878–1967), the Reali School's founder and first principal, was born in Bischofswerda, Saxony. The school was originally established as a preparatory institution for the Technion. Its aim was to impart to its pupils technical knowledge, a theoretical foundation, and a "nationally oriented Hebrew education" (see *Dror 1991*, p. 48).

Two receptions were held in Einstein's honor at the Technion on the morning of 10 February. The first reception was popular in nature. It was attended by approximately fifteen hundred people, including the assistant governor of the northern district, Eric Mills, and the superintendent of police, Northern District,

W. F. Sinclair. The chairman of the Jewish community council in Haifa, Yehuda Eitan, welcomed Einstein on behalf of the community. Council member Lifshitz declared that the council had decided to bestow the title "resident of the Land of Israel" on Einstein. Hillel Jaffe welcomed Einstein on behalf of the Technion Committee (for the French version of his speech, see [AEA, 43 833]). In response to the official greetings, Einstein expressed his admiration for the work of the Jews in this country, whatever their vocation, and "promised to help, as much as he can, in the revival of this land."

The second reception was organized by the Technion Committee and was by invitation only. Einstein was welcomed by Hillel Jaffe, the committee's chairman; Baruch Bina, the representative of the Zionist Commission in Haifa; Shlomo Buzaglo, a member of the Jewish community council; Dr. David Spiegel, a teacher at the Reali School on behalf of the teachers; and Shmuel Pevzner on behalf of the Hadar HaCarmel neighborhood. Einstein also addressed the attendees (see *Do'ar Hayom*, 14 February 1923).

At the time of Einstein's visit, preparations were still being made for introducing a curriculum at the Technion.

246. Rachel-Leah Weizmann-Tchemerinsky (1852?–1939), founder of the first home for the elderly in Haifa and mother of fifteen children, including Chaim Weizmann.

247. The Protestant pastor Martin Schneider (1862–1933), head of the Mount Carmel mission, which was built in 1913. The building in question was most likely the mission itself and not the pastor's home, which had a sloping roof.

248. The Egyptian-born German-language playwright and poet Asis Domet (1890–1943) and his wife Adelheid Domet-Köbke. Domet subsequently wrote to Einstein and asked him whether he remembered being greeted by Domet in Arabic and German and calling him "my Arab friend" in front of a large audience (see Asis Domet to Einstein, 24 September 1929 [AEA, 46 055]).

249. According to the press, the festive banquet took place at the Reali School and not at the Technion. It was attended by the school's principal, teachers, senior pupils, alumni, and distinguished guests, including Shmuel Pevzner, Jaffe, Czerniawski, Elias Auerbach, and Baruch Bina. Auerbach (1882–1971) was a physician and a scholar of Jewish history in biblical times. The school's choir sang for the guests, Arthur Biram and an alumnus gave welcoming speeches for Einstein, and Czerniawski gave a brief talk on relativity. Einstein's speech, which "was steeped in emotion and admiration," was greeted with "wild applause" (see *Do'ar Hayom*, 14 February 1923).

250. According to the press, Einstein visited the Reali School on the morning of 11 February. For his visit to the Technion, see note 245. He toured the Reali's dining room and the workrooms for mechanics, carpentry, and bookbinding. At the end of the visit, Einstein and Elsa planted a tree in the courtyard between the Technion and the Reali School (see *Do'ar Hayom*, 14 February 1923; *The Palestine Weekly*, 16 February 1923). The Russian-Jewish industrialist Michael Polak had forged

close ties with the Rothschild family that led to the establishment of the Portland Cement Syndicate in 1919. The Nesher factory for the production of cement and related products was founded by the syndicate in Yagur, outside Haifa, in 1922. It supplied most of the building products for the *Yishuv*.

The Shemen oil factory, a plant for oil pressing and soap production, was founded in 1920 in Haifa by the Russian-Jewish industrial engineer Nachum Wilbushevitz.

251. The Sea of Galilee.

252. Nahalal was established in 1921 as the first *moshav ovdim* (workers' cooperative settlement). On their tour, the Einsteins were accompanied by Solomon Ginzberg. The moshav's school welcomed the guests. While drinking tea, Einstein chatted with the moshav's council members about the working conditions and about the differences between the moshav ovdim's system and that of other settlement models. They then toured the moshav, and Einstein expressed his amazement that the residents "take care of their beasts much more than they take care of themselves and their family members" (see *Ha'aretz*, 20 February 1923). Richard Kaufmann (1877–1958), architect and urban planner. Nahalal was the first settlement in Palestine he had designed.

253. Migdal lies on the western shore of the Sea of Galilee in the Ginossar Valley north of Tiberias. It was founded as an estate by a group of Zionists from Moscow in 1910. At the time of Einstein's visit, it was in the process of being liquidated. The estate was run by Moshe Glikin (1874–1973), who resided in Haifa at the time. According to press reports, Einstein planted two trees during his visit (see *Regev 2006*, p. 111, and *The Palestine Weekly*, 2 March 1923).

254. The press reported that in Tiberias, Einstein "was accorded a warm welcome by the whole [Jewish] community." Due to "torrents of rain," the plan for Einstein to plant two trees in the new Jewish suburb of Kiryat Shmuel was canceled (see *The Palestine Weekly*, 2 March 1923). It is not clear whether Einstein visited Degania Aleph, the first kibbutz, which was founded in 1909; adjacent Degania Beth, which was founded in 1920; or both. Both lie on the shores of the Sea of Galilee. The Arab village of Mejdal (Magdala) lay on the western shore of the Sea of Galilee. On his return to Tiberias following his tour of the Sea of Galilee settlements, Einstein met with the moderate Mufti Sheikh Taher el Tabari "and other notables of the different communities" (see *The Palestine Weekly*, 2 March 1923).

255. Most likely the Hotel Germania.

256. This was a popular lecture held at the Lämel School in Jerusalem. The lecture was organized by the associations of Jewish doctors, teachers, engineers, and architects in Palestine; the Hebrew Technical Society; and the Palestine Oriental Society. Einstein was introduced to the audience by Yitzhak Ladizhansky, a mathematics teacher at the Hebrew Gymnasia and at the teachers' seminary in Jerusalem, and was greeted by wild applause. His lecture lasted one and a half hours and dealt with the main points of the theory of relativity. It was attended by "the entire

Jerusalem intelligentsia." Prominent members of the audience included Lady Beatrice Miriam Samuel-Franklin and Hadassah Samuel-Grasovsky, Menachem Ussishkin, David Yellin, Judah L. Magnes, the Hebrew philologist Aharon Meir Masie, the historian and Hebrew literature scholar Joseph Klausner, the director of the Department of Education in the Zionist Executive Joseph Lurie, the writer and publicist Mordechai Ben Hillel Hacohen, Father Dhorme, Boris Schatz, "and some of the best minds of the various nations residing in Jerusalem." Approximately 450 people attended the lecture. The chief rabbis were invited but did not attend (see *Do'ar Hayom*, 15 February 1923, and *Ha'aretz* and *The Palestine Weekly*, 16 February 1923).

There was clearly some annoyance in German diplomatic circles that the invitations "did not bear one single European (let alone German) letter, but were only printed in Hebrew." In spite of this criticism, the lecture was presented as a great success: "The crush was incredibly large, so that already 15 minutes before time the gates had to be closed. The audience was mixed: Englishmen, French, Americans etc., Catholics, Protestants, Templars, and for the most part: Jews. It was the first time since the war that Jerusalem had seen such a large gathering that had come to see a German professor at his German lecture" (see *Pressekorrespondenz des Deutschen Auslands-Instituts Stuttgart*, 21 March 1923 [GyBPAAA/R64677]). The diploma of the Jewish Medical Association in Palestine was presented to Einstein by Abraham Albert Ticho (1883–1960), Jerusalem ophthalmologist (see *Ha'aretz*, 16 February 1923).

257. The Einsteins had returned to Government House on 13 February (see *The Palestine Weekly*, 16 February 1923).

258. Frederick H. Kisch referred to the departure in his diary: "Saw Einstein off at Jerusalem station; asked him to let us know if during his tour he had observed that we were doing anything which in his opinion we should not do, or if we were leaving undone things which should be done. He answered: '*Ramassez plus d'argent*' ['collect more money']" (see *Kisch 1938*, p. 31).

259. Max (Mordechai) Mouschly (1874?–1950?), merchant, Jewish community leader, and Zionist functionary in Port Said (see *Ne'eman 2001*, p. 31).

260. Ferdinand de Lesseps (1805–1894) was a French diplomat who oversaw the construction of the Suez Canal. His statue stood at the entrance to the Canal in Port Said.

261. Celia Mouschly-Turkel (1875?–1960). For Einstein's note expressing his gratitude to the Mouschlys for their care of Elsa, see Einstein to Mr. and Mrs. Mouschly, 14/15 February 1923 [*CPAE 2012*, Vol. 13, Abs. 521].

262. The RMS *Ormuz* of the Orient Line. The ship was originally built in 1914 as the S.S. *Zeppelin* for the Norddeutscher Lloyd shipping company. It was purchased and renamed by the Orient Line in 1920 and sailed the route between London and Brisbane. In 1927, it was returned to the ownership of the Norddeutscher Lloyd and renamed the S.S. *Dresden*. In 1934, on its maiden voyage as a cruise ship for the

Nazi "Kraft durch Freude" ("Strength through Joy") leisure organization, it ran aground off the Norwegian island of Karmøy and was eventually scrapped.

263. According to German press reports, there were widespread imprisonments and convictions by the French occupation forces of civil servants, policemen, and businessmen in the Ruhr (see *Berliner Tageblatt*, 16 February 1923, morning edition).

264. A similar fate befell the five pieces of luggage the Einsteins left on board the S.S. *Haruna Maru* in Port Said. The purser informed them that their baggage was denied landing at Marseille by the French customs officials. They were therefore shipped to Amsterdam (see Nippon Yusen Kaisha to Einstein, 26 February 1923 [*CPAE 2012*, Vol. 13, Abs. 528]).

265. Einstein had disembarked at Toulon and traveled by train to Barcelona via Marseille. Press reports stated that he had sent confirmation of his planned arrival in Spain while he was in Singapore, yet he had not given prior notification of the exact time. Therefore, he was not met at the station in Barcelona when he arrived in the evening (see Einstein to Esteve Terradas, 23/24 February 1923 [*CPAE 2012*, Vol. 13, Abs. 527] and *Roca Rossell 2005*, p. 29). According to the German consul-general in Barcelona, the definitive confirmation of his visit was only received one day prior to his arrival and did not reveal the exact date. There are contradictory reports in the Catalan press in regard to Einstein's initial stay in Barcelona. According to one account, he proceeded to the home of Esteve Terradas prior to arriving at Hotel Colón (see *La Veu de Catalunya*, 24 February 1923). In contrast, another account claims that he proceeded to a modest pension, the Cuatro Naciones. The proprietor allegedly attempted to convince him to stay at the Ritz Hotel, where a room had been reserved for him (see *El Debate*, 25 February 1923; Ulrich von Hassell to German Foreign Ministry, 26 February 1923 [GyBPAAA/R 64 677]; and *Sallent del Colombo and Roca Rossell 2005*, p. 74).

Esteve Terradas (1883–1950) was professor of acoustics and optics at the University of Barcelona, member of the Barcelona Royal Academy of Sciences and the Arts, founder of the science section of the Institut d'Estudis Catalans, and one of the earliest disseminators of the theory of special relativity in Spain (see *Glick 1988*, pp. 32–38, and *Roca Rossell 2005*, p. 28). Rafael Campalans (1887–1933) was director of the Industrial School of Barcelona and a Catalan syndicalist politician and engineer by training. Casimiro Lana-Sarrate (1892–?), chemist at the Institute for Electricity and Applied Mechanics in Barcelona. Presumably Ilse von Hassell-Tirpitz (1885–1982), wife of Ulrich von Hassell (1881–1944), the German consul-general in Barcelona and daughter of Admiral Alfred von Tirpitz (1849–1930).

The Refectorium was a restaurant on the Rambla del Centre, part of Barcelona's main pedestrian boulevard, and was frequently patronized by Catalan nationalist politicians (see *Glick 1988*, p. 117).

While he was in Barcelona, Einstein delivered a series of three lectures on relativity at the Institut d'Estudis Catalans in Barcelona. The lectures were held at the Sala d'Actes of the Diputació, the provincial government building, and were sponsored by the Mancomunitat, the Catalan regional authority. The cost of tickets to the lectures was 25 pesetas each. Catalan nationalist symbols were prominently displayed in the lecture hall. The first and second lectures were directed at a scientifically educated audience, the third was intended only for experts. The audience gave Einstein "an extraordinary warm welcome and thanked him with loud applause" (see Ulrich von Hassell to German Foreign Ministry, 26 February 1923 [GyBPAAA/R 64 677]). On 24 February, the first lecture was held at 7 p.m. and dealt with special relativity. The lecture hall was filled to capacity, and there was not sufficient seating for all the invitees (see *La Veu de Catalunya*, 20 February 1923; *La Vanguardia*, 28 February 1923).

On 25 February, the Einsteins visited the Cistercian Romanesque Monastery of Poblet, approximately 80 kilometers west of Barcelona. They were accompanied by Bernat Lassaleta i Perrin, professor of chemistry at the Industrial School; Ventura Gassol, a Catalan writer and nationalist politician; and others. Einstein signed his name in the guest book. He also toured the nearby town of L'Espluga de Francolí (see *Glick 1988*, p. 117). For a photo from the trip, see Illustration 27.

On 26 February, Einstein toured the city of Terrassa, approximately 30 kilometers northwest of Barcelona and home to a famous basilica. He was accompanied by the president of the Mancomunitat of Catalonia, the architect Josep Puig i Cadafalch. At 5 p.m., he paid a formal visit to Valentin Carulla, rector of the University of Barcelona, accompanied by Terradas; university secretary Carlos Calleja y Borja-Tarrius; professor of chemistry Simon Vila Vendrell; and professor of physics Eduardo Alcobe, who was also the president of the Royal Academy (see *Glick 1988*, pp. 117–118). He also received a visitor from the Sociedad de Atracción de Forasteros (Society of Tourist Attractions), who presented him with an illustrated publication on Barcelona. At 7 p.m., he gave his second lecture in the relativity series, on general relativity, to a crowded auditorium. Following the lecture, he had a private meal at the Ritz Hotel with Josep Puig i Cadafalch. Acting Mayor Enric Maynés, and possibly also Campalans, attended the dinner as well (see *Roca Rossell 2005*, p. 30, and *Sallent del Colombo and Roca Rossell 2005*, p. 74).

On 27 February, Einstein toured two innovative schools, the Escola del Mar, an experimental school for physically disabled children, established in 1922, and the Grupo Escolar "Baixeras." At noon, a reception was held at the Consell de Cent del Ayuntamiento (the Barcelona City Hall). Einstein was officially welcomed by Acting Mayor Enric Maynés in Catalan and granted the status "illustrious guest." The mayor praised Einstein's scientific genius and his ethics and pacifism. In his reply, Einstein thanked the mayor for the city's warm welcome and expressed his pleasure that the mayor's speech revealed a desire for an improvement of the political and national discourse (see *La Veu de Catalunya*, 28 February 1923, morning edition). According to another report, he wished for Barcelona "a new human community that would overcome every political and personal rancor" (*Diario de Barcelona*, 28 February 1923; English translation from *Glick 1988*, p. 113; and Ulrich von

Hassell to German Foreign Ministry, 26 February 1923 [GyBPAAA/R 64 677]). In the evening, Einstein lectured at the Real Academia de Ciencias y Artes de Barcelona (the Royal Academy of Sciences and Arts of Barcelona) on the philosophical consequences of relativity and the cosmological implications of a finite universe. The audience was more limited than at the more popular lectures. One of the attendees was Josep Comas Solà, an astronomer and opponent of relativity, who visibly showed his discomfort with Einstein's lecture (see *El Debate*, 2 March 1922). On 6 March, Einstein was nominated as a corresponding member of the physical sciences section of the Academy by Bernat Lassaleta i Perrin, the mathematician Ferran Tallada, and the physicists Ramon Jardí and Tomàs Escriche i Mieg (for the official nomination, see ES-BaACA, "Prof. Einstein y la Reial Acadèmia de Ciències i Arts de Barcelona," 6 March 1923). Following the lecture, Einstein received a delegation from the anarcho-syndicalist Confederación Nacional del Trabajo (CNT) at the Ritz Hotel. They accompanied him to their headquarters in the Baixa de Sant Pere. The delegation included two prominent leaders of the CNT, Angel Pestaña and Joaquín Maurin. Pestaña introduced Einstein at the meeting. Einstein expressed his surprise at the high degree of illiteracy in Spain (which had been cited by Pestaña), and stated that he believed that repression was caused by stupidity rather than by evil. He urged members of the working class to read Spinoza. Some reports also claimed that Einstein remarked to Pestaña: "I too am a revolutionary, but in the area of science. I am concerned with social questions, as are other scientists, because they constitute one of the most interesting aspects of human life" (see *El Diluvio* and *El Noticiero Universal*, 28 February 1923, and *Glick 1988*, pp. 108−109). This quote was widely disseminated in the Spanish and international press. However, Einstein strongly denied the utterance in a subsequent interview with a reporter for the Spanish newspaper *ABC*, stating: "I said that I am not a revolutionary, not even in the scientific area . . ." (See *ABC*, 2 March 1923; *Glick 1988*, pp. 109−112; and *Turrión Berges 2005*, p. 47). In the evening, Campalans hosted a farewell banquet for the Einsteins. Prominent attendees were Terradas and the German-speaking Catalan nationalist politician Miquel Vidal i Guardiola. The menu was written in "relativistic Latin" and contained references to Einstein's theories and to other physicists who were thought to have paved the way for relativity (see *La Publicitat*, 28 February 1923; *Glick 1988*, pp. 120−121; and *Sallent del Colombo and Roca Rossell 2005*, p. 72).

On 28 February, Einstein visited the Escola Industrial de Barcelona as a guest of a school with a clear socialist agenda that fostered education and technology. His host was its director, Rafael Campalans. Einstein witnessed a performance of the *sardana*, the Catalan national dance, by the La Penya de la Dansa troupe and was given records, presumably of music for sardanas. He then toured the port of Barcelona. At 7 p.m., he held his third lecture in the relativity series, which dealt with current problems in relativity (see *La Veu de Catalunya*, 1 March 1923; *Glick 1988*, pp. 119−120; and *Roca Rossell 2005*, p. 30).

266. Following the entry for 22–28 February, Einstein left one complete page and eighteen lines on the next page blank.

267. Esteve Terradas. Ulrich von Hassell and his wife Ilse von Hassell-Tirpitz.

Einstein departed Barcelona by train on the morning of 1 March. According to Hassell's reports, Einstein appeared in Barcelona "always as a German, not as a Swiss" (see Ulrich von Hassell to German Foreign Ministry, 26 February 1923 [GyBPAAA/R 64 677]). Einstein arrived at the Mediodía station in Madrid at 11:30 p.m. on 1 March. A large crowd turned out to greet him. He was met by two official delegations, one from the Universitad Central de Madrid's science faculty, and the other from the College of Physicians. The university's delegation was headed by Blas Cabrera (1878–1945), professor of electricity and magnetism and director of the Laboratory of Physical Research at the Palace of Industry and the Arts. Other members of the delegation were the astronomer Pedro Carrasco and the mathematicians Francisco Vera and Josep Maria Plans. The physicians' delegation was led by the anatomist Julián Calleja. He was also met by the German ambassador, members of the German community, and members of the press. After brief introductions, Einstein left for the Palace Hotel with Julius (Julio) Kocherthaler (?–1927) and his wife, Lina Kocherthaler-Edenfeld. Julius was a co-founder of the General Spanish Mining Company and a distant relative of both Einstein and Fritz Haber.

On the morning of 2 March, the Kocherthalers took the Einsteins for a sightseeing drive through Madrid. Einstein spent the day with Cabrera at his Laboratory of Physical Research. In the evening, the Einsteins attended a musical review titled *Tierra de nadie* at the Teatro Apolo (see *El Debate*, 2 and 3 March 1923; *La Voz* and *La Vanguardia*, 3 March 1923; Ernst Langwerth von Simmern to German Foreign Ministry, 19 March 1923 [GyBPAAA/R 64 677]; and *Glick 1988*, pp. 123–124).

268. On 3 March, Einstein toured the Prado for the first of three visits. Afterward, he was welcomed by the mayor of Madrid, Joaquín Ruiz-Giménez, at City Hall. Einstein delivered all three lectures at the Universidad Central in the physics auditorium. All the lectures "were extraordinarily well-attended." He delivered his first lecture at 6 p.m. The topic of the lecture was special relativity (see "Lectures at the University of Madrid," [*CPAE 2012*, Vol. 13, Appendix H]). It was attended by mathematicians, physicists, philosophers, and politicians, including Antonio Maura, the former prime minister; Amalio Gimeno, the former foreign minister; and Joaquín Salvatella, the minister of public instruction. Pedro Carrasco introduced Einstein to the audience. The lecture was followed by a banquet at the Palace Hotel hosted by the College of Physicians. The dinner was organized by the president of the college, Ignacio Bauer, and its founder, Toribio Zúñiga. It was attended by José Rodríguez Carracido, president of the Academy of Sciences, and prominent Madrid physicians, including Angel Pulido, who campaigned on behalf of Sephardic Jewry (see *El Debate* and *El Liberal*, 4 March 1923; Ernst Langwerth von Simmern to German Foreign Ministry, 19 March 1923 [GyBPAAA/R 64677];

and *Glick 1988*, pp. 124–126). Bauer was also president of the Spanish Federation of the Keren Hayesod, which planned to hold a reception in Einstein's honor (see Secretary of the Zionist Organisation's Executive Committee to M. L. Ortega, 16 March 1923 [IsJCZA/KH1/193]).

269. King Alfonso XIII (1886–1941). José Rodríguez Carracido (1856–1928), president of the Royal Academy of Exact, Physical and Natural Sciences and rector of the Universidad Central. Carracido spoke briefly about the three-tiered structure of science. He claimed that the theory of relativity was an example of the highest tier, that of pure theory (for the transcript of his address, see *Discursos 1923*, pp. 23–25). Notable attendees were Joaquín Salvatella, Ignacio Bauer, the mathematicians Cecilio Jiménez Rueda and Eduardo Torroja, the engineers Leonardo Torres y Quevedo and Nicolás de Ugarte, the geologist Eduardo Hernández Pacheco, and the zoologist Ignacio Bolívar (see *ABC, El Imparcial,* and *El Sol*, 6 March 1923; and *Glick 1988*, pp. 126–127). Cabrera presented an assessment of Einstein's scientific achievements (for a transcript of his lecture, see *Discursos 1923*, pp. 7–15). For Einstein's reply, see *Einstein 1923a*. Following Einstein's address, King Alfonso presented a diploma to Einstein that confirmed his status as a corresponding member of the academy. For the diploma, see "Diploma of the Spanish Academy of Sciences," 4 March 1923 [*CPAE 2012*, Vol. 13, Abs. 531]. In his address, Salvatella offered Einstein "Spain's hospitality and the financial support of the government in case the current conditions in his homeland should make the continuation of his research temporarily impossible!" (see Ernst Langwerth von Simmern to German Foreign Ministry, 19 March 1923 [GyBPAAA/R64 677], and *Glick 1988*, pp. 126–129). For a photo of the occasion, see Illustration 28.

270. A "tea of honor" was hosted by the Marquesa de Villavieja, Doña Petronilla de Salamanca y Hurtado de Zaldívar (1869–1951). Many members of the Madrid intelligentsia and aristocracy were in attendance, including Blas Cabrera, José Rodríguez Carracido, and Joaquín Salvatella; the philosophers José Ortega y Gasset (1883–1955) and Manuel García Morente; the authors Miguel Asúa, José María Salaverría, and Ramón Gómez de la Serna; the neurologists Gonzalo R. Lafora and José M. Sacristán; the physician and scientist Gregorio Marañon, the German paleontologist Hugo Obermaier; and the Vizconde de Eza, Luis de Marchalar y Monreal, a trustee of the Spanish Board for the Advancement of Research. At the event, Einstein and the violinist Antonio Fernández Bordas improvised an "intimate concert" (see *ABC*, 6 and 10 March 1923, and *Glick 1988*, pp. 129–131).

271. Einstein probably used the term "Catholic" here to mean "ascetic."

272. In the afternoon, a special session of the Mathematical Society was held. For a description of its deliberations, see *Glick 1988*, pp. 132–134. Kuno Kocherthaler (1881–1944) was Einstein's distant cousin, co-founder of the General Spanish Mining Company, and an art collector. At 8:30 p.m. Einstein visited Santiago Ramón y Cajal (1852–1934), histologist, psychologist, and Nobel Prize winner. Einstein's second Madrid lecture dealt with general relativity and was held at the

Universidad Central (see *El Imparcial*, 6 March 1923; *El Liberal*, 8 March 1923; and *Glick 1988*, pp. 135–136). For the text of the lecture, see "Lectures at the University of Madrid," [*CPAE 2012*, Vol. 13, Appendix H]. Presumably Wilhelm (Guillermo) Vogel, an associate at the Spanish-German Bank.

273. On the trip to Toledo, the Einsteins were accompanied by Julius (Julio) and Lina Kocherthaler; Kuno Kocherthaler and his wife, the art historian María Luisa Cazurla; Ortega y Gasset; and the art historian Manuel B. Cossío, who was presumably the guide. They toured the Hospital de Santa Cruz; the Plaza de Zocodover; the Cathedral of Saint Mary of Toledo; the medieval Tránsito and Santa María la Blanca synagogues; the Tagus River; and the church of Santo Tomé, in which they viewed El Greco's *Burial of the Count of Orgaz* (see *ABC*, 7 March 1923; *Glick 1988*, pp. 136–138; and Ortega's description in *La Nación*, 15 April 1923).

274. Einstein was accompanied by Carracido at the audience, which took place at the Palacio Real at noon. The Queen Mother was Maria Christina of Austria (1858–1929). For Einstein's invitation to the audience, see Juan Falcó y Trivulzio, Marques de Castel-Rodrigo to Einstein, 6 March 1923 [*CPAE 2012*, Vol. 13, Abs. 534]. The queen consort, Victoria Eugenie of Battenberg, was visiting her mother in Algeciras in the south of Spain and was therefore absent. Earlier in the day, a group of engineering students met with Einstein and invited him to lecture to the Alumni Association of Engineers and Architects. He promised to do so the next day (see *ABC*, 8 March 1923, and *Glick 1988*, p. 138).

Einstein's third lecture on problems raised by the theory of relativity and his work on a unified field theory was held at the Universidad Central. For the text of the lecture, see "Lectures at the University of Madrid," [*CPAE 2012*, Vol. 13, Appendix H]. One journalist estimated that not even one-fifth of the audience understood the lecture. High-ranking representatives of the military were present, including the engineers Emilio Herrera and Joaquín de La Llave (see *El Debate, El Imparcial*, and *El Liberal*, 8 March 1923; and *Glick 1988*, pp. 138–139). The German ambassador was Ernst Langwerth von Simmern (1865–1942). For the invitation, see Ernst Langwerth von Simmern to Einstein, 3 March 1923 [*CPAE 2012*, Vol. 13, Abs. 530]. The reception was held at the German Embassy. It was attended by 110 guests, including Carracido; Blas Cabrera; Manuel García Morente; the educator María de Meatzu; and many physicians, among them Florestán Aguilar, Julián Calleja, Teófilo Hernando, Gustavo Pittaluga, and Sebastiá Recasens (the dean of the Faculty of Medicine at the Universidad Central); as well as members of the German community. The ambassador's wife was Margarete von Simmern-Rottenburg, and his daughter was Juliane von Simmern (1910–?). "No foreign scholar has received such an enthusiastic and extraordinary reception in the Spanish capital in living memory" (see *ABC*, 8 March 1923; Ernst Langwerth von Simmern to German Foreign Ministry, 19 March 1923 [GyBPAAA/R 64 677]; and *Glick 1988*, p. 139).

275. The honorary doctorate was awarded by the Universidad Central de Madrid at a traditional ceremony that commenced at 11 a.m. For the diploma, see Diploma of

honorary doctorate, 8 March 1923 [*CPAE 2012*, Vol. 13, Abs. 539]. See Illustration 30. First, Josep Maria Plans read a biography of Einstein. For the text of Einstein's short address, see "Honorary Doctorate Speech at the University of Madrid," 8 March 1923 [*CPAE 2012*, Vol. 13, Appendix I]. This was followed by presentations by several students. The ambassador gave a speech in Spanish on the history of cultural relations between Germany and Spain (for the text of the speech, see GyBAr (B)/Band 501, German Embassy Madrid, Vorgang Einstein. See also Ernst Langwerth von Simmern to German Foreign Ministry, 19 March 1923 [GyBPAAA/R 64 677]; *Glick 1988*, p. 140; and *Sánchez Ron and Romero de Pablos 2005*, p. 65).

At 12:30 p.m., Einstein visited the Alumni Association of Engineers and Architects, which was the alumni association of the Catholic Institute of Arts and Industries. Einstein gave a brief address in French on the finite nature of the universe (see *ABC* and *El Noticiero*, 9 March 1923, and *Glick 1988*, pp. 141–142). His talk was attended by the deputy Spanish minister for economic development. Einstein was elected an honorary member of the association (for the diploma, see Asociación de Alumnos de Ingenieros y Arquitectos de España to Einstein, 5 March 1923 [*CPAE 2012*, Vol. 13, Abs. 532]). At 6 p.m., Einstein's fourth lecture was held at the Madrid Athenaeum, a literary-scientific club and national university. Its topic was the philosophical consequences of relativity. The event was presided over by the endocrinologist Gregorio Marañon. Einstein was introduced by the marine biologist Odón de Buen, who suggested that their guest lead a joint Spanish-Mexican scientific expedition to observe the upcoming solar eclipse in Mexico (see *La Voz*, 9 March 1923, and *Glick 1988*, pp. 142–143). For Einstein's speech, see *ABC* and *El Heraldo de Madrid*, 9 March 1923, and *Glick 1988*, pp. 143–144. The director of the Madrid Conservatory was Antonio Fernández Bordas (1870–1950).

276. The Einsteins toured the historical royal palace and monastery El Escorial, approximately 45 kilometers northwest of Madrid, and the Castle of the Mendoza in the town of Manzanares el Real, approximately 60 kilometers northwest of Madrid. At 6 p.m., Einstein attended a public tribute in his honor at the Residencia de Estudiantes, a residential college at the Universidad Central. In his remarks, Ortega y Gasset expounded on Einstein's role in the context of science in Western culture and compared him to Newton and Galileo. He also stated that he viewed relativity as the seed of a new culture (see *El Sol*, 10 March 1923; *Glick 1988*, pp. 144, 161–163; and *Sánchez Ron and Romero de Pablos 2005*, p. 53). In his reply, Einstein tried to play down the significance of his theoretical innovations, stating that he was "more of a traditionalist than an innovator." He also allegedly stated that "relativity had not changed anything but had reconciled facts that were irreconcilable by the usual methods" (see *El Sol*, 10 March 1923).

277. Diego Velázquez (1599–1660). El Greco (1541–1614). Wilhelm Ullmann, director of the Deutsche Bank in Madrid, and Thyra Ullmann-Ekwall (1881–1982), a Swedish-born painter. For the invitation to lunch with Langwerth von Simmern, see Ernst Langwerth von Simmern to Einstein, 3 March 1923 [*CPAE 2012*, Vol. 13, Abs. 530].

278. Francisco Goya (1746–1828), Raffaello Santi (1483–1520), and Fra Angelico (c. 1395–1455).

279. At 3 p.m., Einstein was met at the station in Zaragoza by a delegation headed by University of Zaragoza physicist Jerónimo Vecino, who had initiated the visit. Also in attendance were the university's rector, Ricardo Royo-Villanova; the university's secretary-general, Inocencio Jiménez; the chemist Antonio de Gregorio-Rocasolano y Turmo; and professors of the faculty of medicine; the German consul, Gustav Freudenthal, and his daughter; Mayor Basilio Ferrández Milagro; and the chief of public works, Miguel Mantecón. On his arrival, Einstein was driven in the mayor's car to the Hotel Universo-Cuatro Naciones (see *Sánchez Ron and Romero de Pablos 2005*, p. 119).

Einstein gave two lectures in Zaragoza. Both were held in French in the auditorium of the Faculty of Medicine and Sciences. The first lecture, on special relativity, was held on 12 March at 6 p.m. The hall was filled to capacity. Following the lecture, Rocasolano expressed his admiration for Einstein and for the research being carried out in Zaragoza based on Einstein's work. Lorenzo Pardo, secretary of the Zaragoza Academy of Exact Sciences, then awarded him the title of corresponding member.

For the diploma, see Diploma of the Academia de ciencias exactas, fisico-quimicas y naturales de Zaragoza, 13 March 1923 [*CPAE 2012*, Vol. 13, Abs. 544] (see Illustration 32). In the evening, a dinner was held in Einstein's honor at the German consulate.

On 13 March, Einstein had originally planned to give his second lecture in Zaragoza at 11:30 a.m. He then planned to travel to Bilbao to lecture at the Society for Basque Studies. However, the lecture in Bilbao was canceled, and the second Zaragoza lecture was rescheduled for 6 p.m. In the morning of 13 March, Einstein therefore had time to tour Zaragoza. He visited the Basilica-Cathedral of our Lady of the Pillar, the La Seo Cathedral, the Lonja, the medieval commercial exchange, and the Aljafería Palace in the morning. Lunch was held at 1 p.m. at the Centro Mercantile, to which a distinguished group of university professors was invited by the Academy of Sciences. The philologist Domingo Miral hosted the event and gave a brief speech, in which he praised Einstein. In his reply, Einstein remarked that "up to the present moment, he had perceived the throb of the Spanish soul only in Zaragoza" (see *El Heraldo de Aragon*, 14 March 1923). It was also reported that "in Barcelona and in Madrid he experienced the charm of our art that expresses our personalities; but that it was in Zaragoza where, by admiring the architectural monuments, he had found a robust and eloquent expression of our regional physiognomy" (see *El Noticiero*, 14 March 1923). The topic of the second lecture, titled "Space and Time," was general relativity. There were far fewer audience members present. He was introduced by the dean of the science faculty, Gonzalo Calamita. In his lecture, Einstein highlighted the geometrical character of general relativity and some of its experimental evidence. He also discussed attempts to unify electricity and gravitation (see *Sánchez Ron and Romero de Pablos 2005*, p. 122). An honorary certificate was presented to Einstein by the faculty of sciences (see University of Zaragoza to Einstein, 13 March 1923 [*CPAE 2012*, Vol. 13, Abs. 545]). Following the lecture, a banquet was held in Einstein's honor at the German consul's residence. Einstein played the violin. After the meal, Einstein, Vecino, and the consul attended an operetta at the Teatro Principal titled *La Viejecita* ("The Little Old Lady").

On 14 March, his forty-fourth birthday, in the morning, Einstein toured Rocasolano's laboratory, who was conducting research on Brownian motion in colloids, and the university classrooms. He bid farewell to various government and university officials. He then had lunch at the Hotel Universo with the German pianist Emil Sauer, who was also visiting Zaragoza. During dessert, a dance troupe performed a traditional Spanish dance, the *jota*, for which Einstein expressed enthusiasm. Thereafter he departed Zaragoza by train for Barcelona, where he spent another day without any public obligations before leaving for Zurich (see *El Heraldo de Aragon*, 13–15 March 1923; *El Noticiero*, 14 March 1923; Pilger to German Foreign Ministry, 21 March 1923 [GyBPAAA/R 64677]; *Glick 1988*, pp. 145–149; and *Sánchez Ron and Romero de Pablos 2005*, p. 125).

# Additional Texts

1. TLS. [AEA, 36 423]. Published in *CPAE 2012*, Vol. 13, Doc. 21, pp. 86–87. Attached to Jun Ishiwara to Einstein, 26 January 1922 [*CPAE 2012*, Vol. 13, Doc. 40].

2. Deletions and addition of "September" and "October" in Einstein's hand.

3. TDS (IsJCZA, A222/165).

4. The original invitation is not extant.

5. Present-day Jakarta (see text of diary, this volume, notes 49 and 200).

6. Einstein seems to have envisaged a trip to Palestine as early as the autumn of 1921 (see Chaim Weizmann to Einstein, 7 October 1921 [*CPAE 2009*, Vol. 12, Doc. 259]).

7. The Hebrew University of Jerusalem.

8. REPT. Published on 3 November 1922 in *The Straits Times*, 3 November 1922, and as Appendix D in *CPAE 2012*, Vol. 13, p. 853. The speech was delivered at Belle Vue, the home of Manasseh Meyer, on 2 November 1922.

9. See text of diary, this volume, entry for 2 November 1922.

10. A reference to the Jewish people.

11. A reference to the Hebrew University of Jerusalem.

12. A reference to the *numerus clausus*, the restrictions on the admission of Jewish students at Eastern European universities (see *Motta 2013*, p. 53).

13. REPT. Published under the title "Plauderei über meine Eindrücke in Japan" in *Kaizo*, January 1923, pp. 343–338. Also published in *CPAE 2012*, Vol. 13, Doc. 391, pp. 605–612. A manuscript (Dr. Hiroshi Miyake, Kobe University) [EPPA, 71 716]

is also available, written on letterhead "The Kanaya Hotel Nikko, Japan." Einstein visited Nikko on 7 December 1922.

14. Dated by the reference to this article in text of diary, this volume, entry for 7 December 1922.

15. Sanehiko Yamamoto was the president of the Kaizo-Sha publishing house. For his invitation, see Text 1 in the Additional Texts section of this volume.

16. He arrived on 17 November 1922.

17. ALS. [AEA, 36 430]. Published in *CPAE 2012*, Vol. 13, Doc. 397, pp. 617–618. Written on letterhead "The Miyako Hotel, Kyoto."

18. The S.S. *Haruna Maru*.

19. Morikatsu Inagaki, who was serving as Einstein's personal interpreter. Tony Inagaki.

20. The function was a banquet hosted by the Japanese-German Society at the Osaka Hotel (see text of diary, this volume, entry for 11 December 1922, note 143).

21. For Einstein's description in his diary of Elsa's reaction, see text of diary, this volume, entry for 11 December 1922.

22. ALSX. [EPPA, 75 620]. Published in *CPAE 2012*, Vol. 13, Doc. 400, p. 642. Written on letterhead "The Miyako Hotel, Kyoto." Hans Albert (1904–1973) and Eduard (1910–1965) Einstein.

23. At the Swiss Polytechnic in Zurich. He registered as a first-semester student in engineering on 8 September 1922 (see "Matrikel für Einstein, Albert, von Zürich, geb. 14. Mai 1904" [SzZuETH, Diplomarchiv]).

24. Hans Albert had asked his father to take him on the trip (see Hans Albert Einstein to Einstein, after 24 June 1922 [*CPAE 2012*, Vol. 13, Doc. 246]).

25. According to Einstein and Mileva's divorce decree, if Einstein were to be awarded the Nobel Prize, the prize money would be deposited in a Swiss bank as Mileva's property, less 40,000 marks. The interest would be entirely at her disposal, but she could draw on the capital only with Einstein's consent (see Divorce Decree, 14 February 1919 [*CPAE 2004*, Vol. 9, Doc. 6]).

26. Apparently Mileva intended to buy a house with the capital. Eventually, she purchased a house at Huttenstrasse 62 in Zurich for 105,000 Swiss francs, which was roughly equivalent to 19,125 U.S. dollars (see "Kaufvertrag," 26 May 1924; SzZuZB/Heinrich Zangger Estate).

27. To attend the next meeting of the International Committee on Intellectual Cooperation, of which Einstein was a member, which was scheduled for the spring of 1923 (see Einstein to Max Wertheimer, 18 September 1922 [*CPAE 2012*, Vol. 13, Doc. 362]).

28. Nickname for Eduard.

29. Mileva Einstein-Marić.

30. TTrL. (GyBSA, I. HA, Rep.76 Vc, Sekt. 1, Tit. XI, Teil Vc, Nr. 55, Bl. 158). Published in *Steinberg et al. 1967*, p. 269; *Grundmann 2004*, p. 233, and *CPAE 2012*, Vol. 13, Doc. 402, p. 643. Solf was the German ambassador to Tokyo.

31. Dated by Solf. This letter is contained in Solf's report to the German Foreign Ministry, 3 January 1923 [GyBSA, I. HA, Rep. 76 Vc, Sekt. 1, Tit. 11, Teil 5c, Nr. 55, Bl. 157–158].

32. In the text of diary, this volume, Einstein had mentioned that the matter was too complicated for a telegram (see entry for 20 December 1922).

33. For the background to Maximilian Harden's criticism of Einstein's absence from Germany, see text of diary, this volume, note 170.

34. ALSX. Published in *Ishiwara 1923*, n. pag., and *CPAE 2012*, Vol. 13, Doc. 405, p. 645. [EPPA, 92 817]. Ishiwara had served as Einstein's interpreter at his scientific lectures in Tokyo, and they had collaborated on the electromagnetic problem of the general theory of relativity during Einstein's tour of Japan (see *CPAE 2012*, Vol. 13, Introduction, p. lxxiv).

35. Dated by the days on which Einstein stayed in Moji (see text of diary, this volume).

36. ALS (JSeTU). Published in *Doi, B. 1932*, pp. 11–14, and *CPAE 2012*, Vol. 13, Doc. 411, pp. 655–656. [EPPA, 90 965]. Torn. On stationery "Nippon Yusen Kaisha S.S. 'Haruna Maru.'"

37. "To the Great Einstein," before 30 December 1922 [*CPAE 2012*, Vol. 13, Abs. 486].

38. In his poem, Doi praises Einstein's "immortal name" and compares his "genius" to "a newly emerging comet" (see *Doi, B. 1932*, pp. 5–6).

39. In his poem, Doi criticizes what he perceives as Japan's "insularity," its "imitation of [the West's] external technology," and its "lagging behind [the West] by a hundred years" (see *Doi, B. 1932*, pp. 6–7).

40. Doi had given Einstein an album of woodprints by Hokusai on 3 December (see text of diary, this volume).

41. *Einstein 1922–1924*. For Einstein's preface, see *Einstein 1923c*.

42. Eiichi Tsuchii (Doi). See Text 10 in the Additional Texts section of this volume.

43. ALS (JSeTU). [EPPA, 90 964]. Published in *CPAE 2012*, Vol. 13, Doc. 412, p. 657. Torn. The envelope is addressed "Herrn Bansui Tsuchii 21 Moto-Aramachi Sendai (Japan)," and postmarked "Shanghai 3 Jan 1923." Eiichi (1909–1933) was the son of Bansui Tsuchii.

44. Dated by the reference to this document in Text 9 in the Additional Texts section in this volume.

45. ALSX. [AEA, 122 794]. Published in *CPAE 2012*, Vol. 13, Doc. 414, p. 658. Yoshi Yamamoto was the wife of Sanehiko Yamamoto.

46. Dated on the assumption that this document is mentioned in text of diary, this volume, entry for 30 December 1922.

47. Misako and Sayoko Yamamoto.

48. REPT. Published in *The China Press*, 3 January 1923, and in *CPAE 2012*, Vol. 13, Appendix F, p. 858. The speech was delivered at the reception of the Quest Club in Shanghai on 1 January 1923 and published in *The China Press*, 3 January 1923.

49. The Hebrew University of Jerusalem.

50. While his years at the Zurich Polytechnic were, to some extent, a moratorium in the evolvement of his Jewish identity, Einstein had forged a strong sense of his Jewishness during his school years in Munich (see *Rosenkranz 2011*, pp. 14–29).

51. For Einstein's analysis of the plight of German Jewry, see "Assimilation and Anti-Semitism," 3 April 1920, and "Anti-Semitism. Defense through Knowledge," after 3 April 1920 [*CPAE 2002*, Vol. 7, Docs. 34 and 35].

52. On the evolvement of Einstein's affiliation with the Zionist movement, see *Rosenkranz 2011*, pp. 46–85.

53. ALS (SSVA, Svante Arrhenius Archive, Letters to Svante Arrhenius, vol. E1:6). Published in *CPAE 2012*, Vol. 13, Doc. 420, pp. 697. [EPPA, 73 210]. Written on stationery "Nippon Yusen Kaisha S.S. 'Haruna Maru.'" Arrhenius (1859–1927) was the head of the Nobel Institute for Physical Chemistry and acting chairman of the Nobel Committee for Physics.

54. According to Chinese press reports, Einstein received a cable with the news after he arrived in Shanghai on 13 November (see text of diary, this volume, note 50). A day later, Ilse Einstein informed Christopher Aurivillius, secretary of the Royal Swedish Academy of Sciences, that she had forwarded the information about the committee's decision to Einstein by letter (see Ilse Einstein to Christopher Aurivillius, 14 November 1922 [*CPAE 2012*, Vol. 13, Abs. 446]).

55. Margarete Hamburger (1869–1941) was a German-Jewish philosopher and Berlin admirer of Einstein.

56. The Nobel Foundation deposited 121,572.54 Swedish kronor to an account for Einstein at a bank in Stockholm. According to the contemporary exchange rate, this amount was equivalent to $32,654 (see Hendrik Sederholm and Knut A. Posse to Einstein, 11 December 1923 [*CPAE 2012*, Vol. 13, Doc. 396]).

57. Niels Bohr (1885–1962) was professor of theoretical physics at the University of Copenhagen; he received the Nobel Prize for Physics for the year 1922.

58. ALS (DkKoNBA). Published in *Bohr 1977*, p. 686, and *CPAE 2012*, Vol. 13, Doc. 421, pp. 697–698. [EPPA, 89 896]. Written on stationery "Nippon Yusen Kaisha S.S. 'Haruna Maru.'"

59. Niels Bohr to Einstein, 11 November 1922 [*CPAE 2012*, Vol. 13, Doc. 386].

60. Einstein had left Japan on 29 December 1922 (see text of diary, this volume).

61. Probably *Bohr 1922*.

62. He worked on the problem on board the S.S. *Haruna Maru* from 30 December on (see text of diary, this volume), and completed the manuscript of *Einstein 1923b* in January.

63. Arthur Stanley Eddington (1882–1944) was professor of astronomy and experimental philosophy at the University of Cambridge and director of its observatory. Hermann Weyl (1885–1955) was professor of mathematics at the Swiss Polytechnic in Zurich.

64. PLS. Published in *Kaizo*, February 1923, pp. 195–196, *Kaneko 1987*, p. 377, and *CPAE 2012*, Vol. 13, Doc. 426, pp. 714–715. The German original is unavailable. The

Japanese Proletarian Alliance was a radical left-wing group that had been influenced by the French anti-war and internationalist Clarté movement.

65. Dated by the original published in *Kaizo*, February 1923, pp. 195–196.

66. The Alliance had asked Einstein the following questions: "1. What are your views on the ----- imperialist government of Japan? 2. What do you hope for Japanese youth?" (see Japanese Proletarian Alliance to Einstein, 12 December 1922 [*CPAE 2012*, Vol. 13, Abs. 471]). The censored word may have been "aggressive" (see *Kaneko 1987*, p. 368).

67. Sanehiko Yamamoto.

68. The Washington Naval Conference, which was attended by representatives from nine nations with interests in the Western Pacific, among them Japan, had ended on 6 February 1922 with the Washington Naval Treaty, which set limits to Japan's development of its fleet and settled questions of its territorial claims.

69. AKS. [AEA, 124 316]. Greetings by Hannah Ruppin omitted. Not stamped or postmarked. Addressed to "Mr. Dr. Arthur Ruppin." Written on verso of drawing of the S.S. *Belgenland*, which includes a sketch by Einstein of himself and Hannah Ruppin (see Illustration 34).

70. Dated by the days on which Einstein met Hannah Ruppin (see text of diary, this volume, entries for 3 and 5 February 1923).

71. Hannah Ruppin-Hacohen (1892–1985).

72. At the time of Einstein's visit, Ruppin was raising funds in the United States for a mortgage bank and other Zionist financial institutions (see *Wasserstein 1977*, p. 272, note 3).

73. REPT. Published in *Jüdische Rundschau* 33 (24 April 1923): 195–196. First published in Hebrew (see *Einstein 1923d*). For the English version, see *Einstein 1923f*. Also published in *CPAE 2015*, Vol. 14, Doc. 15, pp. 46–49. This translation is reprinted from *The New Palestine* 4 (1923): 341.

74. Einstein toured Palestine between 2 and 14 February 1923 at the invitation of the Zionist Organisation's Palestine Bureau (see text of diary, this volume).

75. Einstein visited Tel Aviv 8–9 February 1923. During his tour, he expressed his admiration for the lively development of the city (see text of diary, this volume, entry for 8 February 1923).

76. Einstein visited the "Bezalel" Art Academy in Jerusalem on 6 February 1923. A day earlier, he toured two neighborhoods that were being constructed by workers' cooperatives (see text of diary, this volume, entries for 5 and 6 February 1923).

77. For subsequent correspondence regarding the loan interest, see Julius Simon to Einstein, 29 June 1923 [*CPAE 2015*, Vol. 14, Abs. 111].

78. Einstein noted the hardships faced by the members of Kibbutz Degania, which he visited on 12 February 1923 (see text of diary, this volume).

79. Einstein met with Arab-German poet Asis Domet on 10 February 1923 in Haifa, and two days later with moderate Arab notables in Tiberias (see text of diary, this volume, entry for 10 February 1923 and note 254).

80. The expert in question may have been the Sephardi physician Angel Pulido, whom Einstein had met in Madrid in March 1923 (see text of diary, this volume, note 268).

81. See the text of diary, this volume, entry for 12 February 1923.

82. Richard Kaufmann (1877–1958) was a German-Jewish architect and urban planner.

83. The British Mandatory Authority.

84. Einstein visited the Technion on 10 and 11 February 1923 (see text of diary, this volume).

85. Einstein gave a lecture on relativity at the future site of the Hebrew University on Mount Scopus on 7 February 1923 (see text of diary, this volume). For Einstein's fund-raising efforts on behalf of the American Jewish Physicians' Committee, see *CPAE 2009*, Vol. 12, Introduction, p. xxxiv.

86. Einstein edited the first volume dedicated to mathematics and physics (see *Scripta Universitatis atque Bibliothecae Hierosolymitanarum. Mathematica et Physica* 1 [1923]).

# References

## Newspapers and Periodicals

*ABC* (Madrid)
*Aspeklarya* (Jerusalem)
*Berliner Tageblatt*
*The China Press* (Shanghai)
*El Debate* (Madrid)
*Diario de Barcelona*
*El Diluvio* (Barcelona)
*Do'ar Hayom* (Jerusalem)
*Fukuoka Nichinichi Shinbun*
*Ha'aretz* (Tel Aviv)
*Ha'Olam* (Jerusalem)
*El Heraldo de Aragon* (Zaragoza)
*El Heraldo de Madrid*
*Hinode Shinbun* (Kyoto)
*El Imparcial* (Madrid)
*Israel's Messenger* (Shanghai)
*Japan Times & Mail* (Tokyo)
*Jüdische Presszentrale Zürich*
*Jüdische Rundschau* (Berlin)
*Kahoku Shimpo* (Sendai)
*Kaizo* (Tokyo)
*Latest News and Wires through Jewish Correspondence Bureau News and Telegraphic Agency* (New York)
*El Liberal* (Madrid)
*Min Guo Ri Bao* (Canton)

*La Nación* (Madrid)
*The New Palestine* (New York)
*El Noticiero* (Zaragoza)
*El Noticiero Universal* (Barcelona)
*Osaka Mainichi Shinbun*
*The Palestine Weekly* (Jerusalem)
*Pressekorrespondenz des Deutschen Auslands-Instituts Stuttgart*
*La Publicitat* (Barcelona)
*Scripta Universitatis atque Bibliothecae Hierosolymitanarum. Mathematica et Physica* (Jerusalem)
*Shin Aichi* (Nagoya)
*El Sol* (Madrid)
*South China Morning Post* (Hong Kong)
*The Straits Times* (Singapore)
*Taisho 11 nen Nisshi Tokyo-joshi-koto-shihan-gakkou* (Tokyo)
*Tokyo Asahi Shinbun*
*Tokyo Nichinichi Shinbun*
*La Vanguardia* (Barcelona)
*La Veu de Catalunya* (Barcelona)
*La Voz* (Madrid)
*Waseda Gakuho* (Tokyo)
*Yomiuri Shinbun* (Tokyo)
*Zionistische Korrespondenz* (Berlin)

## Books and Articles

*Aichi 1923*   Aichi, Keichi. "Kogishitsu ni okeru Einstein [Einstein in the Lecture Room]." *Kaizo*, February 1923, pp. 299–301.

*Aschheim 1982*   Aschheim, Steven E. *Brothers and Strangers. The East European Jew in German and German Jewish Consciousness, 1800–1923.* Madison: University of Wisconsin Press, 1982.

*Bailey 1989*   Bailey, Herbert Smith, Jr. "On the Collected Papers of Albert Einstein: The Development of the Project." *Proceedings of the American Philosophical Society* 133, no. 3 (1989): 347–359.

*Bellah 1972*   Bellah, Robert N. "Intellectual and Society in Japan." *Daedalus* 101, no. 2 (Spring 1972): 89–115.

*Ben-Arieh 1989*   Ben-Arieh, Yehoshua. "Perceptions and Images of the Holy Land." In *The Land That Became Israel. Studies in Historical Geography*, ed. Ruth Kark, pp. 37–53. New Haven: Yale University Press, 1989.

*Bergman 1919*   Bergman, Hugo S. "ha-Ve'ida ha-Universitayit." *Ha'Olam* 11 (26 December 1919): 3–6.

*Bergman 1974*   ———. "Personal Remembrance of Albert Einstein." In *Logical and Epistemological Studies in Contemporary Physics*, ed. Robert S. Cohen and Marx W. Wartofsky, pp. 388–394. Dordrecht: Reidel, 1974.

*Bergson 1922*   Bergson, Henri. *Durée et simultanéité, à propos de la théorie d'Einstein*. Paris: Alcan, 1922.

*Berkowitz 2012*   Berkowitz, Michael. "The Origins of Zionist Tourism in Mandate Palestine: Impressions (and Pointed Advice) from the West." *Public Archeology* 11, no. 4 (2012): 217–234.

*Bohr 1922*   Bohr, Niels. "Der Bau der Atome und die physikalischen und chemischen Eigenschaften der Elemente." *Zeitschrift für Physik* 9 (1922): 1–67.

*Bohr 1977*   ———. *Collected Works*. Vol. 4, *The Periodic System (1920–1923)*, ed. J. Rud Nielsen. Amsterdam: North-Holland, 1977.

*Butenhoff 1999*   Butenhoff, Linda. *Social Movements and Political Reform in Hong Kong*. Westport, CT: Praeger, 1999.

*Calaprice 2015 et al.*   Calaprice, Alice, Daniel Kennefick, and Robert Schulmann, eds. *An Einstein Encyclopedia*. Princeton, NJ: Princeton University Press, 2015.

*Clifford 2001*   Clifford, Nicholas R. "The Long March of 'Orientalism': Western Travelers in Modern China." *New England Review* 22, no. 2 (2001): 128–140.

*Confino 2003*   Confino, Alon. "Dissonance, Normality, and the Historical Method. Why Did Some Germans Think of Tourism after May 8, 1945?" In *Life after Death. Approaches to a Cultural and Social History of Europe during the 1940s and 1950s*, ed. Richard Bessel and Dirk Schumann, pp. 323–347. Cambridge: Cambridge University Press, 2003.

*CPAE 1987*   Einstein, Albert. *The Collected Papers of Albert Einstein*. Vol. 1, *The Early Years, 1879–1902*, ed. John Stachel et al. Princeton, NJ: Princeton University Press, 1987.

*CPAE 1993*   ———. *The Collected Papers of Albert Einstein*. Vol. 5, *The Swiss Years: Correspondence, 1902–1914*, ed. Martin J. Klein et al. Princeton, NJ: Princeton University Press, 1993.

*CPAE 1998*   ———. *The Collected Papers of Albert Einstein*. Vol. 8, *The Berlin Years: Correspondence, 1914–1918*, ed. Robert Schulmann et al. Princeton, NJ: Princeton University Press, 1998.

*CPAE 2002*   ———. *The Collected Papers of Albert Einstein*. Vol. 7, *The Berlin Years: Writings, 1918–1921*, ed. Michel Janssen et al. Princeton, NJ: Princeton University Press, 2002.

CPAE 2004 ———. *The Collected Papers of Albert Einstein*. Vol. 9, *The Berlin Years: Correspondence, January 1919–April 1920*, ed. Diana Kormos Buchwald et al. Princeton, NJ: Princeton University Press, 2004.

CPAE 2006 ———. *The Collected Papers of Albert Einstein*. Vol. 10, *The Berlin Years: Correspondence, May–December 1920 and Supplementary Correspondence, 1909–1920*, ed. Diana Kormos Buchwald et al. Princeton, NJ: Princeton University Press, 2006.

CPAE 2009 ———. *The Collected Papers of Albert Einstein*. Vol. 12, *The Berlin Years: Correspondence, January–December 1921*, ed. Diana Kormos Buchwald et al. Princeton, NJ: Princeton University Press, 2009.

CPAE 2012 ———. *The Collected Papers of Albert Einstein*. Vol. 13, *The Berlin Years: Writings & Correspondence, January 1922–March 1923*, ed. Diana Kormos Buchwald et al. Princeton, NJ: Princeton University Press, 2012.

CPAE 2015 ———. *The Collected Papers of Albert Einstein*. Vol. 14, *The Berlin Years: Writings & Correspondence, April 1923–May 1925*, ed. Diana Kormos Buchwald et al. Princeton, NJ: Princeton University Press, 2015.

CPAE 2018 ———. *The Collected Papers of Albert Einstein*. Vol. 15, *The Berlin Years: Writings & Correspondence, June 1925–May 1927*, ed. Diana Kormos Buchwald et al. Princeton, NJ: Princeton University Press, 2018.

Craig 2014 Craig, Claudia. "Notions of Japaneseness in Western Interpretations of Japanese Garden Design, 1870s–1930s." *New Voices* 6 (January 2014): 1–25.

Dirlik 1996 Dirlik, Arif. "Chinese History and the Question of Orientalism." *History and Theory* 35, no. 4 (December 1996): 96–118.

Discursos 1923 Real Academia de Ciencias Exactas, Fìsicas y Naturales. *Discursos pronunciados en la sesión solenne que se dignó presidir S. M. el Rey el dia 4 de marzo de 1923, celebrada para hacer entrega del diploma de académico corresponsal al profesor Alberto Einstein*. Madrid: Talleres Poligráficos, 1923.

Doi 1922 Doi, Uzumi. *Einstein sotaiseiriron no hitei* [*Rebuttal of Einstein's Theory of Relativity*]. Tokyo: Sobunkan, 1922.

Doi, B. 1932 Doi, Bansui, ed. *Ajia ni sakebu* [*Crying in Asia*]. Tokyo: Hakubun-Sha, 1932.

Doron 1980 Doron, Joachim. "Rassenbewusstsein und naturwissenschaftliches Denken im deutschen Zionismus während der wilhelminischen Ära." *Jahrbuch des Instituts für Deutsche Geschichte* 9 (1980): 389–427.

Dror 1991 Dror, Yuval. "The Hebrew Technion in Haifa, Israel (1902–1950): Academic and National Dilemmas." *History of Higher Education Annual* 11 (1991): 45–60.

Eddington 1920 Eddington, Arthur S. *Space Time and Gravitation: An Outline of the General Relativity Theory*. Cambridge: Cambridge University Press, 1920.

Eddington 1921 ———. "A Generalization of Weyl's Theory of the Electromagnetic and Gravitational Fields." *Royal Society of London, Proceedings* A99 (1921): 104–122.

Einstein 1922–1924 Einstein, Albert. *Einstein Zenshu* [*The Collected Works of Einstein*]. Jun Ishiwara et al., trans. Tokyo: Kaizo-Sha, 1922–1924.

Einstein 1923a ———. [To the Spanish Academy of Sciences.] *Discursos*, 1923, pp. 19–20.

*Einstein 1923b* ———. "Zur allgemeinen Relativitätstheorie." *Preußische Akademie der Wissenschaften* (Berlin). *Physikalisch-mathematische Klasse. Sitzungsberichte* (1923): 32–38.

*Einstein 1923c* ———. "Vorwort." In *Einstein 1922–1924*, vol. 2, pp. [i–ii].

*Einstein 1923d* ———. "Impressions from My Trip to Palestine" (in Hebrew). *Ha'Olam* 11, no. 14 (20 April 1923): 269.

*Einstein 1923e* ———. "Prof. Einstein über seine Eindrücke in Palästina." *Jüdische Rundschau* 33 (1923): 195–196.

*Einstein 1923f* ———. "My Impressions of Palestine." *The New Palestine* 4 (11 May 1923): 341. (English translation of *Einstein 1923e*.)

*Eisinger 2011* Eisinger, Josef. *Einstein on the Road*. Amherst, NY: Prometheus Books, 2011.

*Eliav 1976* Eliav, Binyamin, ed. *HaYishuv BeYamei HaBait Ha Leumi*. Jerusalem: Keter, 1976.

*Ezawa 2005* Ezawa, Hiroshi. "Impacts of Einstein's Visit on Physics in Japan." *AAPPS Bulletin* 15, no. 2 (2005): 3–16.

*Falk 2006* Falk, Rafael. "Zionism, Race, and Eugenics." In *Jewish Tradition and the Challenge of Darwinism*, ed. Geoffrey Cantor and Marc Swetlitz, pp. 137–162. Chicago: University of Chicago Press, 2006.

*Fischer 2003* Fischer, Conan. *The Ruhr Crisis, 1923–1924*. Oxford: Oxford University Press, 2003.

*Foster 1982* Foster, Stephen W. "The Exotic as a Symbolic System." *Dialectical Anthropology* 7, no.1 (1982): 21–30.

*Friedman 1977* Friedman, Menachem. *Society and Religion: The Non-Zionist Orthodox in Eretz-Israel 1918–1936*. Jerusalem: Yad Yitzhak Ben-Zvi Publications, 1977 (Hebrew).

*Friedman 2001* Friedman, Robert M. *The Politics of Excellence: Behind the Nobel Prize in Science*. New York: Holt, 2001.

*Fuhrmann 2011* Fuhrmann, Malte. "Germany's Adventures in the Orient. A History of Ambivalent Semicolonial Entanglements." In *German Colonialism: Race, the Holocaust, and Postwar Germany*, ed. Volker Langbehn and Mohammed Salama, pp. 123–145. New York: Columbia University Press, 2011.

*Gelber 2000* Gelber, Mark H. *Melancholy Pride. Nation, Race, and Gender in the German Literarture of Cultural Zionism*. Tübingen: Max Niemeyer, 2000.

*Germana 2010* Germana, Nicholas A. "Self-Othering in German Orientalism: The Case of Friedrich Schlegel." *The Comparatist* 34 (May 2010): 80–94.

*Ginsburg 2014* Ginsburg, Lisa. "Worlds Apart in Singapore. A Jewish Family Story." *Asian Jewish Life* no. 15 (October 2014): 20–29.

*Glick 1987* Glick, Thomas, ed. *The Comparative Reception of Relativity*. Dordrecht: Reidel, 1987.

*Glick 1988* ———. *Einstein in Spain: Relativity and the Recovery of Science*. Princeton, NJ: Princeton University Press, 1988.

*Goldstein 1980* Goldstein, Yaakov. "Were the Arabs Overlooked by the Zionists?" *Forum on the Jewish People, Zionism and Israel* 39 (1980): 15–30.

*Gordon 2003* Gordon, Andrew. *The Modern History of Japan*. Oxford: Oxford University Press, 2003.

Grundmann 2004    Grundmann, Siegfried. *Einsteins Akte. Wissenschaft und Politik—Einsteins Berliner Zeit*, 2nd ed. Berlin: Springer, 2004.

Hambrock 2003    Hambrock, Matthias. *Die Etablierung der Außenseiter. Der Verband nationaldeutscher Juden 1921–1935*. Cologne: Böhlau Verlag, 2003.

Hashimoto 2005    Hashimoto, Yorimitsu. "Japanese Tea Party: Representations of Victorian Paradise and Playground in *The Geisha* (1896)." In *Histories of Tourism: Representation, Identity, and Conflict*, ed. John K. Walton, pp. 104–124. Clevedon, UK: Channel View Publications, 2005.

Hiki 2009    Hiki, Sumiko. *Einstein karano bohimei* [*The Epitaph from Einstein*]. Tokyo: Demado-Sha, 2009.

Hu 2005    Hu, Danian. *China and Albert Einstein: The Reception of the Physicist and His Theory in China, 1917–1976*. Cambridge, MA: Harvard University Press, 2005.

Hu 2007    ———. "The Reception of Relativity in China." *Isis* 98 (2007): 539–557.

Inagaki 1923a    Inagaki, Morikatsu. "Einstein hakase ni otomo site [Having Accompanied Dr. Einstein]." *Josei Kaizo* 2, no. 2 (1923):176–187.

Inagaki 1923b    ———. "Einstein hakase no nihon-ongaku-kan [Dr. Einstein's View of Japanese Music]." *Josei Kaizo* 2, no. 1 (1923): 113–115.

Isaacson 2008    Isaacson, Walter. *Einstein. His Life and Universe*. London: Pocket Books, 2008.

Ishiwara 1923    Ishiwara, Jun. *Einstein kyoju koen-roku* [*Records of Professor Einstein's Lectures*]. Tokyo: Kaizo-Sha, 1923.

Jackson 1992    Jackson, Anna. "Imagining Japan: The Victorian Perception and Acquisition of Japanese Culture." *Journal of Design History* 5, no. 4 (1992): 245–256.

Jansen 1989    Jansen, Marius B. "Einstein in Japan." *Princeton University Library Chronicle* 50, no. 2 (1989): 145–154.

Johnson 1993    Johnson, Sheila K. "Review: *The Jews and the Japanese: The Successful Outsiders*, by Ben-Ami Shiloni." *Monumenta Nipponica* 48, no. 1 (Spring 1993): 136–139.

Jokinen and Veijola 1997    Jokinen, Eeva, and Soile Veijola. "The Disoriented Tourist: The Figuration of the Tourist in Contemporary Cultural Critique." In *Touring Cultures*, ed. Chris Rojek and John Urry, pp. 23–51. London: Routledge, 1997.

Kagawa 1920    Kagawa, Toyohiko. *Seishin undo to shakai undo* [*Spiritual and Social Movement*]. Tokyo: Keishi-Sha Shoten, 1920.

Kaiser 1992    Kaiser, Wolf. "The Zionist Project in the Palestine Travel Writings of German-Speaking Jews," *Leo Baeck Institute Yearbook* 37 (1992): 261–286.

Kalland and Asquith 1997    Kalland, Arne, and Pamela J. Asquith. "Japanese Perceptions of Nature. Ideas and Illusions." In *Japanese Images of Nature. Cultural Perspectives*, ed. Pamela J. Asquith and Arne Kalland, pp. 1–35. London: RoutledgeCurzon, 1997.

Kaneko 1981    Kaneko, Tsutomu. *Einstein Shock to Taisho Era*, 2 vols. Tokyo: Kawade Shobo Shinsha, 1981.

Kaneko 1984    ———. "Einstein's View of Japanese Culture." *Historia Scientiarum* 27 (1984): 51–76.

Kaneko 1987    ———. "Einstein's Impact on Japanese Intellectuals: The Socio-Cultural Aspects of the 'Homological Phenomena.'" In *Glick 1987*, pp. 351–379.

*Kaneko 2005* ———. "Einstein's Impact on Japanese Culture." *AAPPS Bulletin* 15, no. 6 (2005): 12–17.

*Kaplan 1997* Kaplan, E. Ann. *Looking for the Other. Feminism, Film, and the Imperial Gaze.* New York: Routledge, 1997.

*Kark and Oren-Nordheim 2001* Kark, Ruth, and Michal Oren-Nordheim. *Jerusalem and Environs. Quarters, Neighborhoods, Villages.* Jerusalem and Detroit: Magnes Press and Wayne State University Press, 2001.

*Keitz 1993* Keitz, Christine. "Die Anfänge des Massentourismus in der Weimarer Republik." *Archiv für Sozialgeschichte* 33 (1993): 179–209.

*Kisch 1938* Kisch, Frederick H. *Palestine Diary.* London: Gollancz, 1938.

*Koshar 1998* Koshar, Rudy. "'What Ought to Be Seen.' Tourists' Guidebooks and National Identities in Modern Germany and Europe." *Journal of Conttemporary History* 33, no. 3 (July 1998): 323–340.

*Kotzebue 1792* Kotzebue, August von. *Vom Adel.* Leipzig: Paul Gotthelf Kummer, 1792.

*Kretschmer 1921* Kretschmer, Ernst. *Körperbau und Charakter. Untersuchungen zum Konstitutions-Problem und zur Lehre von den Temperamenten.* Berlin: Springer, 1921.

*Krobb 2014* Krobb, Florian. "'Welch' unbebautes und riesengroßes Feld': Turkey as Colonial Space in German World War I Writings." *German Studies Review* 37, no. 1 (February 2014): 1–18.

*Kuwaki 1934* Kuwaki, Ayao. *Einstein-den* [*The Biography of Einstein*]. Tokyo: Kaizo-Sha, 1934.

*Lambourne 2005* Lambourne, Lionel. *Japonisme: Cultural Crossings between Japan and the West.* Berlin: Phaidon, 2005.

*Lary 2006* Lary, Diana. "Edward Said: Orientalism and Occidentalism." *Journal of the Canadian Historical Association* 17, no. 2 (2006): 3–15.

*Leerssen 2000* Leerssen, Joep. "The Rhetoric of National Character. A Programmatic Survey." *Poetics Today* 21, no. 2 (2000): 267–292.

*Lipphardt 2016* Lipphardt, Veronika. "The Emancipatory Power of Heredity: Anthropological Discourse and Jewish Integration in Germany, 1892–1935." In *Heredity Explored. Between Public Domain and Experimental Science, 1850–1930*, ed. Staffan Müller-Wille and Christina Brandt, pp. 111–139. Cambridge, MA: MIT Press, 2016.

*Lissak 1993* Lissak, Moshe, ed. *Toldot HaYishuv HaYehudi B'Eretz Yisrael, Me'az Ha-Aliyah HaRishona. Tkufat HaMandat HaBriti.* Jerusalem: Israel Academy of Sciences, 1993.

*Long 2006* Long, Brian. *Nikon: A Celebration.* Ramsbury, UK: Crowood Press, 2006.

*Lubrich 2004* Lubrich, Oliver. "'Überall Ägypter.' Alexander von Humboldts orientalistischer Blick auf Amerika." *Germanisch-romanische Monatsschrift* 54 (2004): 19–39.

*Malamat et al. 1969* Malamat, Abraham, Haim Hillel Ben-Sasson and Samuel Ettinger, eds. *Toldot Am Yisrael.* Vol. 3, *Toldot Am Yisrael B'Et HaHadasha*, ed. Shmuel Ettinger. Tel Aviv: Dvir, 1969.

*Mansfield 2006–2007* Mansfield, Maria Luisa. "Jerusalem in the 19th Century: From Pilgrims to Tourism." *ARAM Periodical* 19 (2006–2007): 705–714.

*Marchand 2001*   Marchand, Suzanne. "German Orientalism and the Decline of the West." *Proceedings of the American Philosophical Society* 145, no. 4 (December 2001): 465–473.

*Metzler and Wildt 2012*   Metzler, Gabriele, and Michael Wildt, eds. *Über Grenzen. 48. Deutscher Historikertag in Berlin 2010. Berichtsband*. Göttingen: Vandenhoeck & Ruprecht, 2012.

*Miles 1982*   Miles, Robert. *Racism and Migrant Labour. A Critical Text*. London: Routledge and Kegan Paul, 1982.

*Miles and Brown 2003*   Miles, Robert, and Malcolm Brown. *Racism*, 2nd ed. London: Routledge, 2003.

*Motta 2013*   Motta, Giuseppe. *Less Than Nations. Central-Eastern European Minorities after WWI*, Vol. 1. Newcastle upon Tyne, UK: Cambridge Scholars Publishing, 2013.

*Mudimbe-Boyi 1992*   Mudimbe-Boyi, Elisabeth. "Travel, Representation, and Difference, or How Can One Be a Parisian?" *Research in African Literatures Journal* 23 (1992): 25–39.

*Nagashima 1923*   Nagashima, Juetsu. "Einstein hakase shotai-ki [Report on Inviting Dr. Einstein]." *Hitotsubashi* 80 (1923): 136–137.

*Nakamoto 1998*   Nakamoto, Seigyo. *Kanmon Fukuoka no Einstein [Einstein in Kanmon and Fukuoka]*. Shimonoseki: Shin-Nihon-Kyoiku-Tosho, 1998.

*Nathan and Norden 1975*   Nathan, Otto, and Heinz Norden. *Albert Einstein: Über den Frieden. Weltordnung oder Weltuntergang?* Bern: Lang, 1975.

*Ne'eman 2001*   Ne'eman, Yuval. "Al Savta Yocheved, al Einstein ve-al abba [About Grandma Yocheved, Einstein and Father]." *Igeret Ha'Akademia LeMada'im* 21 (November 2001): 31–32.

*Neumann and Neumann 2003*   Neumann, Helga, and Manfred Neumann. *Maximilian Harden (1861–1927). Ein unerschrockener deutsch-jüdischer Kritiker und Publizist*. Würzburg: Königshausen & Neumann, 2003.

*Niewyk 2001*   Niewyk, Donald L. *The Jews in Weimar Germany*. New Brunswick, NJ: Transaction Press, 2001.

*Nisio 1979*   Nisio, Sigeko. "The Transmission of Einstein's Work to Japan." *Japanese Studies in the History of Science* 18 (1979): 1–8.

*Nünning 2008*   Nünning, Ansgar. "Zur mehrfachen Präfiguration/Prämediation der Wirklichkeitsdarstellung im Reisebericht: Grundzüge einer narratologischen Theorie, Typologie und Poetik der Reiseliteratur." In *Points of Arrival: Travels in Time, Space, and Self/Zielpunkte: Unterwegs in Zeit, Raum und Selbst*, ed. Marion Gymnich et al., pp. 11–32. Tübingen: Francke, 2008.

*Okamoto 1923*   Okamoto, Ippei. "Gaku-sei wo e ni shite [Portraying a Master Scholar]." *Kaizo*, February 1923, pp. 140–147.

*Okamoto 1981*   ———. "Albert Einstein in Japan: 1922." *American Journal of Physics* 49 (1981): 930–940.

*Patiniotis and Gavroglu 2012*   Patiniotis, Manolis, and Kostas Gavroglu. "The Sciences in Europe: Transmitting Centers and the Appropriating Peripheries." In *The

*Globalization of Knowledge in History*, ed. Jürgen Renn, pp. 321–343. Berlin: Edition Open Access, 2012.

*Poiger 2005*  Poiger, Uta G. "Imperialism and Empire in Twentieth-Century Germany." *History & Memory* 17, no. 1/2 (Spring/Summer, 2005): 117–143.

*Porat and Shavit 1982*  Porat, Yehoshua, and Ya'akov Shavit, eds. *HaMandat Ve'HaBayit HaLeumi (1917–1947)*. Jerusalem: Keter, 1982.

*Pratt 1985*  Pratt, Mary Louise. "Scratches on the Face of the Country: Or, What Mr. Barrow Saw in the Land of the Bushmen." *Critical Inquiry* 12, no. 1 (Autumn, 1985): 119–143.

*Pratt 1992*  ———. *Imperial Eyes. Travel Writing and Transculturation*. London: Routledge, 1992.

*Regev 2006*  Regev, Yoav. *Migdal: Moshava leChof haKineret [Migdal: Moshava by the Shores of the Sea of Galilee]*. Netanya, Israel: Hotza'at Achiassaf, 2006.

*Renn 2013*  Renn, Jürgen. "Einstein as a Missionary in Science." *Science & Education* 22 (2013): 2569–2591.

*Roca Rossell 2005*  Roca Rossell, Antoni. "Einstein en Barcelona." *Quark: Ciencia, Medicina, Comunicación y Cultura 36* (2005): 26–35.

*Root 2013*  Root, Hilton L. *Dynamics among Nations. The Evolution of Legitimacy and Development in Modern States*. Cambridge, MA: MIT Press, 2013.

*Rosenkranz 1999*  Rosenkranz, Ze'ev. "Albert Einstein's Travel Diary to Palestine, February 1923." *Arkhiyyon: Reader in Archives Studies and Documentation* 10–11 (1999): 184–207 (in Hebrew).

*Rosenkranz 2011*  ———. *Einstein before Israel: Zionist Icon or Iconoclast?* Princeton, NJ: Princeton University Press, 2011.

*Sabrow 1994a*  Sabrow, Martin. "Märtyrer der Republik. Zu den Hintergründen des Mordanschlags am 24. Juni 1922." In *Walther Rathenau: 1867–1922; die Extreme berühren sich; eine Ausstellung des Deutschen Historischen Instituts in Zusammenarbeit mit dem Leo Baeck Institute*, ed. Ursel Berger and Hans Wilderotter, pp. 221–236. New York and Berlin: Argon, 1994.

*Sabrow 1994b*  ———. *Der Rathenaumord. Rekonstruktion einer Verschwörung gegen die Republik von Weimar*. Munich: R. Oldenbourg Verlag, 1994.

*Sabrow 1999*  ———. *Die verdrängte Verschwörung: Der Rathenau-Mord und die deutsche Gegenrevolution*. Frankfurt am Main: Fischer Taschenbuch, 1999.

*Sachs 2003*  Sachs, Aaron. "The Ultimate 'Other': Post-Colonialism and Alexander von Humboldt's Ecological Relationship with Nature." *History and Theory* 42 (December 2003): 111–135.

*Said 1978*  Said, Edward. *Orientalism*. London: Routledge & Kegan Paul, 1978.

*Sallent del Colombo and Roca Rossell 2005*  Sallent del Colombo, Emma, and Antoni Roca Rossell. "La cena 'relativista' de Barcelona (1923)." *Quark: Ciencia, medicina, comunicación y cultura 36* (2005): 72–84.

*Samuel 1945*  Samuel, Herbert Louis Viscount. *Memoirs*. London: Cresset Press, 1945.

*Sánchez Ron and Romero de Pablos 2005*  Sánchez Ron, José M., and Ana Romero de Pablos. *Einstein en España*. Madrid: Publicaciones de la Residencia de Estudiantes, 2005.

*Saposnik 2006*    Saposnik, Arieh B. "Europe and Its Orients in Zionist Culture before the First World War." *Historical Journal* 49, no. 4 (2006): 1105–1123.

*Sayen 1985*    Sayen, Jamie. *Einstein in America. The Scientist's Conscience in the Age of Hitler and Hiroshima*. New York: Crown, 1985.

*Seagrave and Seagrave 1999*    Seagrave, Sterling, and Peggy Seagrave. *The Yamato Dynasty: The Secret History of Japan's Imperial Family*. New York: Broadway Books, 1999.

*Selwyn 1996*    Selwyn, Tom. "Introduction." In *The Tourist Image. Myths and Myth Making in Tourism*, ed. Tom Selwyn, pp. 1–32. Chichester, UK: Wiley, 1996.

*Seth and Knox 2006*    Seth, R., and C. Knox. *Weimar Germany between Two Worlds. The American and Russian Travels of Kisch, Toller, Holtischer, Goldschmidt, and Rundt*. New York: Peter Lang, 2006.

*Shachori 1990*    Shachori, Ilan. *Halom shehafach le-Krach—Tel Aviv, Leida ve-Zmicha [A Dream That Turned into a City—Tel Aviv. Birth and Growth]*. Tel Aviv: Avivim, 1990.

*Steinberg et al. 1967*    Steinberg, Heiner, Anneliese Griese, and Siegfried Grundmann. *Relativitätstheorie und Weltanschauung. Zur politischen und wissenschaftspolitischen Wirkung Albert Einsteins*. Berlin: Deutscher Verlag der Wissenschaften, 1967.

*Sugimoto 2001a*    Sugimoto, Kenji. *Einstein no Tokyo-Daigaku Kogiroku [Record of Einstein's Lectures at the University of Tokyo]*. Tokyo: Ootake-Shuppan, 2001.

*Sugimoto 2001b*    ———. *Einstein Nihon de soutairon wo kataru [Einstein Talks about Relativity in Japan]*. Tokyo: Kodan-Sha, 2001.

*Takahashi 1921–1927*    Takahashi, Yoshio. *Taisho meiki kan [The Catalogue of Excellent Articles (for Tea Ceremonies) in the Taisho Era]*. Tokyo: Taisho Meiki Hensan-Sho, 1921–1927.

*Takahashi 1933*    ———. "Einstein hakase no raian [Dr. Einstein's Visit to My Hut]." In *Hoki no Ato*, pp. 450–454. Tokyo: Syuho-En, 1933.

*Tsuchii 1920*    Tsuchii, Bansui. *Su le orme dell'Ippogrifo [Temba no michi ni]*, trans. H. Shimoi and E. Jenco. Naples: Saukurà, 1920.

*Turrión Berges 2005*    Turrión Berges, Javier. "Einstein en España." *Monografías de la Real Academia de Ciencias de Zaragoza* 27 (2005): 35–68.

*Walton 2009*    Walton, John K. "Histories of Tourism." In *The SAGE Handbook of Tourism Studies*, ed. Tazim Jamal and Mike Robinson, pp. 115–129. London, SAGE, 2009.

*Waseda 2010*    Waseda University. *Waseda Daigaku nyugaku annnai [Waseda University Guide Book]*. Tokyo: Waseda University, 2010.

*Wasserstein 1977*    Wasserstein, Bernard, ed. *The Letters and Papers of Chaim Weizmann*. Vol. 11, Series A: January 1922–July 1923. New Brunswick, NJ: Transaction Books and Rutgers University, 1977.

*Weiss 2006*    Weiss, Yfaat. "Identity and Essentialism. Race, Racism, and the Jews at the Fin de Siècle." In *German History from the Margins*, ed. Neil Gregor, Nils Roemer, and Mark Roseman, pp. 49–68. Bloomington and Indianapolis: Indiana University Press, 2006.

*Weyl 1918*    Weyl, Hermann. "Gravitation und Elektrizität." *Königlich Preußische Akademie der Wissenschaften* (Berlin) *Sitzungsberichte* (1918): 465–478, 478–480. Reprinted in *Weyl 1968*, vol. 2, pp. 29–42.

*Weyl 1968* ———. *Gesammelte Abhandlungen*, ed. K. Chandrasekharan. Berlin: Springer, 1968.

*Wiemann 1995* Wiemann, Volker. "'Das ist die echte orientalische Gastfreundlichkeit.' Zum Konzept kolonisierbarer, nicht-kolonisierbarer und kolonisierender Subjekte bei Karl May." *KultuRRevolution* 32/33 (1995): 99–104.

*Wilke 2011* Wilke, Sabine. "Von angezogenen Affen und angekleideten Männern in Baja California: Zu einer Bewertung der Schriften Alexander von Humboldts aus postkolonialer Sicht." *German Studies Review* 34, no. 2 (May 2011): 287–304.

*Winteler-Einstein 1924* Winteler-Einstein, Maja. "Albert Einstein—Beitrag für sein Lebensbild," 1924. Typescript. Besso Estate, Basel.

*Yamamoto 1934* Yamamoto, Sanehiko. "Kaizo no Jugo-nen [Fifteen Years of Kaizo]." *Kaizo* 16 (April 1934): 134–143.

*Yapp 2003* Yapp, Malcolm E. "Some European Travelers in the Middle East." *Middle Eastern Studies* 39, no. 2 (2003): 211–227.

*Yokozeki 1956* Yokozeki, Aizo. "Omoideno sakka-tachi: Einstein; Nihon ni nokosita episode [Episodes from Einstein's Visit in Japan]." Reprinted in *Omoideno sakka-tachi* [*Writers in My Memory*]. Tokyo: Hosei University Press, 1956.

*Youngs 2013* Youngs, Tim. *The Cambridge Introduction to Travel Writing*. Cambridge: Cambridge University Press, 2013.

# Index

Page numbers in italics refer to illustrations.
Albert Einstein is abbreviated to "AE" in subentries.

Einstein, Albert (*continued*)
on Greeks: in antiquity, 75, 89
Haifa: AE on, 39, 63; AE on the Technion,
      263; visits, 62, 223, 225; visits Reali
      School, 225, 315n245, 316n249, 316n250;
      visits the Technion, 225, 315n245
Hebrew University of Jerusalem: AE on
      academic standards of, 17; AE on need
      for, 245, 256, 264; fundraising for, 111, 115,
      119, 209, 291–292n38, 308n200; lectures
      at future site of, 39, 71, 72, 74, 219, 221,
      311–312n230; plans visit to, 10; research
      on malaria at, 263; Zionists' perception
      of AE's views on, 244
Hong Kong: AE on, 21–22, 42, 48, 56, 63; 1st
      visit, 123, 125, 127, 129, 131; 2nd visit, 195,
      197; AE on Chinese in, 125, 127, 131, 195;
      AE on Chinese on mainland adjacent
      to, 129; AE on colonial rule in, 42; AE
      on Jewish community, 75, 123, 125, 127;
      AE on segregation in, 22, 58; Jewish
      community in, 195, 197, 292n40
humanism, limits to AE's, 43, 76
ideological constructs of, 16, 75
illnesses during trip, 97, 101, 233
Indians: AE on, 18–19, 42, 103, 105, 107, 131,
      199, 201; AE on alleged intellectual
      inferiority of, 19, 58, 105. *See also* AE:
      Colombo; AE: Negombo; Sinhalese,
      AE on
indigenous peoples, AE ashamed about
      his behavior toward, 18, 105
as insider/outsider, 50
International Committee on Intellectual
      Cooperation, AE's membership in, 12,
      328n27
international scientific communities, AE
      on cooperation between, 67–68, 157,
      245, 294n59
introspection, trip as motivation for, 51,
      60, 74
as an introvert, 65

Japan: AE on, 1, 24–32; AE on adoption
      of Western culture by, 32, 48, 251, 255;
      AE on Chinese influence, 157, 175; AE
      on invitation to, 7–8, 245–246, 253;
      AE's concept of J. prior to trip, 25–26;
      arrival in, 141; article on impressions
      of, 171, 245–251; comparison to
      United States, 173; criticism of tour,
      69; departure from, 191; enthusiasm
      for, 258; enthusiasm for AE in, 68;
      exhaustion from tour of, 65, 189, 252;
      farewell message, 307n187; hectic
      nature of travel in, 65; impact of tour,
      69; interview on impressions of, 153;
      invitation to, 5–8, 243; on Japanese
      architecture, 28, 143, 151, 169, 171, 175, 179,
      181, 248–249; on J. art, 31–32, 151, 155,
      157, 177, 181, 248, 250, 254, 298n103; on
      J. culture, 159, 161, 256; on J. education,
      247; on J. food, 249; on J. labor, 70, 259;
      on J. landscapes, 28, 143, 165, 167, 169,
      179, 181, 183, 189, 246, 248; on J. morals,
      30, 157; on J. music, 28, 31, 107, 109, 121,
      147, 153, 155, 157, 159, 163, 249–250; on
      J. overpopulation, 259; on J. patriotism,
      93, 109, 246; on J. politics, 259, 261; on
      J. scholars, 299n110; on the J. scientific
      community, 254; on J. society, 247–248;
      on J. theater, 31, 145, 147, 157, 250; lecture
      tour in, AE on, 252; motivation for trip
      to, 8, 183, 246, 294n59; nature of AE's
      travel in, 61; passes Mt. Fuji, 143; plans
      for lecture tour of, 6–7, 243; political
      aspects of tour, 69–70; rumors that
      AE plans to migrate to, 302n124;
      success of tour, 68–69; on ties with
      Germany, 175; visits Lake Biwa; 179,
      181; visits Matsushima Islands, 165;
      visits Miyajima, 181; visits Nagoya,
      171, *171*; visits Nara, 181; visits Nikko,
      28, 165, 167, 169; visits Shimonoseki,
      185, 189; warns of danger of militarism

Einstein, Albert (*continued*)

Marseille: returns to, 233; AE visits, 81, 83; AE on women in, 47, 49

misogyny of, 22, 58, 129, 131

Moji: AE visits, 65, 185, 189, 251; departs for return voyage, 191; plays violin at children's party, *189*; plays violin at Commercial Club, 189

morals, attitude toward, 75

Naples, AE visits, 233

national character: AE on 16, 23; firm believer in, 53, 75

on nationalism, 53, 75. *See also* AE on Jewish nationalism

Negombo, AE visits, 49, 203, 205

on "negroes," 115, 119, 291n37

Nobel Prize for Physics: AE awarded, 66–67; AE on, 252, 257, 258; AE learns of being awarded, 257, 293n50; arrangements for, 257, 328n25; certificate, *67*, medal, *257*

Orient, the, AE on, 46–47, 121

"Oriental" identity of, 45–46

Orientalism of, 47–48, 75

Orientals, AE on 17, 46–48

Osaka: AE on, 151; AE visits, 63, 173, 252; Japanese-German Society in, 303n143; popular lecture in, 173

Oslo, AE visits, 5

the "Other:" AE on 15–16, 77; projections onto, 51–52

Palestine: AE and his Zionist hosts in, 62–63, 70–71; AE on, 20, 35–39; AE on agrarian settlements in, 37–38, 39, 262; AE on Arab community in, 36–37, 39, 211, 217, 229; AE on architecture of, 37, 263; AE on debt of agrarian workers, 262; AE on Degania, 263; AE on economy of, 264; AE on European influence in, 39, 48; AE fever in, 72; AE on Jewish immigration to, 38–39, 48; AE on *kibbutzim* in, 229; AE on landscapes of, 35, 37, 39, 211, 213, 215, 217, 227, 229; AE on P. as moral center for Diaspora Jewry, 38, 264; AE as national icon in, 72; AE on Nazareth, 48; AE on "Oriental" quality of, 39; AE on revival of Hebrew, 264; AE visits Sea of Galilee, 48, 225, 227; AE's lodgings in, 62; Arab-Jewish conflict in, AE on, 36–37, 225, 262, 311n229; article on impressions of, 261–264; on British Mandate government, 42, 263; confident of success of Jewish settlement in, 262, 264; critical of Jewish and Arab intellectuals in, 262; on enjoying stay in P., 261; entry into, 211; enthusiasm for P., 261–262, 264; glorification of AE in, 71; impact of AE's tour on, 70–72; instrumental importance of P. for AE, 38, 62; invitation to, 9–11, 244; Jaffa, 33–34, 36, 63, 221; on Jewish community in, 35–36, 38, 211, 261, 312n232, 314n238, 315n240, 316n245; on Jewish laborers in, 37, 217, 262; on malaria, 227, 229, 262, 263; meets with Arabs, 36, 317n254; national significance of AE's tour, 70, 71; nature of AE's travel in, 61–63; passes Majdal (Magdala), 229; plans to visit, 9–10, 38; preparations for trip, 62; reception of AE by Jewish community in, 71–72; on role of P. for Jewish diaspora, 38; travels to Lod, 62, 231; urged to settle in, 51, 231, 311n229; visits Ben Shemen, 314n239; visits Degania, 227, 229; visits Jericho, 215, 217; visits Migdal, 227; visits Mikve Israel, 223; visits Nahalal, 225, 227; visits Nazareth, 62, 225, 227, 229; visits Rishon LeZion, 223, *223*; visits Tiberias, 227; visits Zionist Executive, 310n222; welcomed at Lod, 211; Zionist propaganda use of tour, 71, 74. *See also* AE: Haifa; AE: Jerusalem; AE: Tel Aviv

244; in Tel Aviv, *221*; in Tokyo, *7*, *26*, *30*, *31*, *145*, *151*, 163, *165*, 295n68; as traveler, 74; trip to Kyoto, 177; on trip to Java, 308n198

Einstein, Hans Albert, 65, 252–253

Einstein, Ilse, 3, 10, 40, 76

Einstein, Margot, 3, 76

Einstein-Marić, Mileva, 13, 253, 288n308, 328n25

Eitan, Yehuda, 316n245

Elazari-Vulkani, Yitzhak, 313n236

electromagnetic energy tensor for isotropic ponderable matter, AE's work on, 177, 191

Endo, Yoshitoshi, 301n120

Epstein, Paul, 17

El Escorial, AE on, 41, 239

Escriche i Mieg, Tomàs, 321n265

exoticism, 32, 43, 44, 59

"Exposition Universelle" (Paris), 25

Far East, the, 1. *See also* China; Japan

Ferguson-Davie, Charles James, 115, 117

Fernández Bordas, Antonio, 239, 323n270

Ferrández Milagro, Basilio, 326n279

Fra Angelico, 241

Frankel, Abraham, *115*, 199

Frankel, Julian, *115*

Frankel, Rosa, *115*, 199

Frankel, Tila, *115*

Frankel-Clumeck, Marie, *115*

*Frankfurter Zeitung*, 191

Freudenthal, Gustav, 326n279

Friesland, Tsadok van, 310n222

Fukuda, Tokuzo, 296n69

Fujisawa, Rikitaro, 153, 155

Fujita, Toshihiko, 165

García Morente, Manuel, 323n270, 324n274

Gassol, Ventura, 320n265

Gatton, S., 191, 193

Gavroglu, Kostas, 73

German Foreign Ministry: reports on Einstein's visits, 33, 292–293n48, 296n70, 318n256, 322n267

German foreign policy, positive impact of AE's tour on, 70

German Jews, racial ideas of, 54–55, 58

German right-wing extremists, 12; threats on AE's life, 13–15

German Zionists, racial ideas of, 54–55, 58

Germany: colonial past of, 42; French-German reparations negotiations, 308n204; political scene in, 12; tourism in, 59

Gimeno, Amalio, 322n268

Ginsburg, Charles R., *115*

Ginzberg (Ginzburg), Rosa, 11

Ginzberg (Ginzburg), Solomon, 11, 211, 213, 215, 223, 310n222, 310n226, 312n232, 317n252

Glick, Thomas F., 69, 73

Glikin, Moshe, 317n253

Gluck, Christoph W., 159

Gluskin, Ze'ev, 315n240

Gobin (Hong Kong guide), 195, 197, 292n41

Goldstein (Zionist Organisation), 211

Gómez de la Serna, Ramón, 323n270

Goya, Francisco, 241

El Greco, AE on 41, 237, 239

Greenfield, Caroline, 308n200

Gregorio-Rocasolano y Turmo, Antonio de, 326n279, 327n279

Grendi, Edoardo, 59

*Ha'aretz* (newspaper), 72

Haber, Fritz, 11, 322n267

Hagenbeck, Carl, 203

Haldane, Richard B, 139

Hamburger, Margarete, 257

Harden, Maximilian: AE's reaction to, 183, 253; on AE's trip to Japan, 33, 70, 305n170

Hassell, Ulrich von, 325, 319n265